T0202187

Philosophical Papers

Philosophical Papers

Paul Humphreys

OXFORD
UNIVERSITY PRESS

OXFORD
UNIVERSITY PRESS

Oxford University Press is a department of the University of Oxford. It furthers
the University's objective of excellence in research, scholarship, and education
by publishing worldwide. Oxford is a registered trade mark of Oxford University
Press in the UK and certain other countries.

Published in the United States of America by Oxford University Press
198 Madison Avenue, New York, NY 10016, United States of America.

© Oxford University Press 2019

CIP data is on file at the Library of Congress
ISBN 978-0-19-933487-2

1 3 5 7 9 8 6 4 2

Printed by WebCom, Inc., Canada

To Rosemary, Norman, and Mark—sister, brother-in-law, and brother extraordinaires

{ CONTENTS }

{ ACKNOWLEDGMENTS }

Chapter 1. Originally published in *PSA: Proceedings of the Biennial Meeting of the Philosophy of Science Association 1990*, vol. 2, edited by Arthur Fine, Mickey Forbes, and Linda Wessels, 497–506. Chicago: University of Chicago Press for the Philosophy of Science Association. © 1991 by the Philosophy of Science Association. Reprinted with permission.

Chapter 2. Originally published in *Science in the Context of Application*, edited by Martin Carrier and Alfred Nordmann., 131–142. Berlin: Springer. © Springer Science+Business Media B.V. 2008. Reprinted with permission of Springer.

Chapter 3. Originally published in *Synthese* 169 (2009): 615–26. © Springer Science+Business Media B.V. 2008. Reprinted with permission of Springer.

Chapter 4. Originally published in *Patrick Suppes: Scientific Philosopher*, vol. 2, edited by Paul Humphreys., 103-118. Dordrecht: D. Reidel, 1995. © 1994 Kluwer Academic Publishers. Reprinted with permission of Springer.

Chapter 5. Originally published in *Philosophy of Science* 64 (1997): 1–17. © 1997 by the Philosophy of Science Association. Reprinted with permission.

Chapter 6. Originally published in *Philosophy of Science* 64 (1997): S337–45. © 1997 by the Philosophy of Science Association. Reprinted with permission.

Chapter 7. Originally published in *Minds and Machines* 18 (2008): 431–42. © Springer Science+Business Media B.V. 2008. Reprinted with permission of Springer.

Chapter 8. Originally published in *Philosophy of Science* 75 (2008): 584–94. © 2008 by Paul W. Humphreys.

Chapter 9. Originally published in *Philosophical Review* 94 (1985): 557–70. © 1985 Paul Humphreys.

Chapter 10. Originally published in *British Journal for the Philosophy of Science* 55 (2004): 667–80. © British Society for the Philosophy of Science 2004. Reprinted by permission of Oxford University Press.

Chapter 11. Originally published in *Probability and Statistics: Essays in Honor of David A. Freedman*, edited by Terry Speed and Deborah Nolan, 1–11. Beachwood, OH: Institute of Mathematical Statistics Collections, 2008. © Institute of Mathematical Statistics 2008. Reprinted with permission.

Chapter 12. Originally published in *Synthese* 48 (1981): 225–32. © 1981 by D. Reidel Publishing Co., Dordrecht, Holland, and Boston, U.S.A. Reprinted with permission.

Chapter 13. Originally published in *Science, Explanation, and Rationality: The Philosophy of Carl G. Hempel*, edited by James Fetzer., 267-286. Oxford: Oxford University Press, 2001. © 2001 Paul Humphreys.

Chapter 14. Originally published in *Scientific Metaphysics*, edited by Don Ross, James Ladyman, and Harold Kincaid, 51–78. Oxford: Oxford University Press, 2013. © 2013 Paul Humphreys.

Chapter 15. Previously unpublished. © 2017 Paul Humphreys.

Philosophical Papers

Introduction

During a summer break in college, I set out to solo hike the Pennine Way, a 267-mile footpath snaking across the moors of northern England. One day, my first goal was Stoodley Pike, a hill topped by a 121-foot stone folly. That morning, the Pike was covered in a dense fog, with visibility down to about ten yards. I lost the trail, and for about an hour, I had no idea where I was or whether I was hiking toward or away from the monument. I seemed destined to spend the rest of my holiday on the Pike, but suddenly the mist parted and right in front of me was the folly. When doing philosophy I am often reminded of that morning.

Locating oneself geographically and philosophically involves local and nonlocal information. A benefit of rereading one's earlier work is the discovery of overarching themes that are not evident when individual articles are written. The fifteen papers in this collection are grouped in four general categories: the philosophy of computational science, emergence, the philosophy of probability, and scientific metaphysics and epistemology. On the surface, these themes appear to have little in common and indeed, to an extent, forcing connections between them would be unnatural. But all of these papers are motivated by positions in general philosophy of science, nailed to a specific area of scientific activity. Those positions are a scientific realism; the failure of empiricism as an adequate scientific epistemology and the concomitant need to relinquish an anthropocentric view of epistemology; the need to incorporate scientific and mathematical results into one's metaphysical conclusions; philosophy's susceptibility to being misled by ordinary language; and the exaggerations in claims to the effect that we are trapped within a linguistic framework.

These positions are intimately connected with very general questions such as the relation between mathematics and science, what the different modes of understanding in science are, and why scientific methods are epistemically effective. These are all questions in general philosophy of science, an area that has become more difficult to pursue in recent decades because of increasing specialization and the entrenchment of subdisciplines that have their internal research agendas. But it is also better informed than it once was.

The choice of articles was dictated in some cases by citation levels, in others by what seems in retrospect to be novelty of ideas, and in still other cases by personal preference.

The articles are reprinted as they originally appeared except for omitting the occasional encomium, filling in incomplete reference information, making spelling and other points of style consistent, and correcting typographical errors. Each section in this volume is followed by an extended postscript. Rather than construct a separate postscript for each paper, it turned out to be more informative to combine the comments on the material with an eye to capturing the progress that has been made since their original publication.

I.1. Method

An approach to the philosophy of science that is too focused on case studies is philosophically uninformative. It leads to a kind of deforestation problem, with the philosophical trees mostly dead. Yet when we turn to investigations of how science is applied, highly abstract approaches are rarely instructive, omitting as they do most, if not all, of the intricate specifics that are required to bring scientific models into contact with data.[1] Balancing these considerations is not easy and I doubt if anything both true and general can be said about where the right balance lies. When the tension is cast in terms of an "in principle" versus "in practice" choice, we can say something informative. One of the themes running through Part I is the need for philosophy of science to explicitly incorporate the role of technology into its considerations of scientific method, especially in the area of theory and model application. This attention to the details of how science is applied requires a concession from the "in principle" approach that philosophy has often taken. There is a legitimate place for in-principle results when what is at stake is what cannot be known in principle, because what cannot be known in principle cannot be known in practice. But different considerations enter when we are interested in positive results.

Demonstrating that something can be known in principle legitimizes research in that area. But such a claim has a status similar to existence results in mathematics where it is known, for example, that a solution to a particular type of differential equation with generic boundary or initial conditions exists

---

[1] Suppes 1962 is an exception to the above claim, being sensitive to the complexities of bringing together abstract theory and data, although his central example is highly stylized.

but we have no idea what the specific form of that solution looks like. This is undoubtedly interesting but it is only a starting point for further investigation. Within philosophy, however, in-principle claims can spiral off into a nether region in which claims are made that God could have the in-principle knowledge, claims that are epistemologically dubious whatever one's deistic beliefs. Hypothetical omniscient beings would not become scientists; scientific methods would instead be exhibits in a Museum of Epistemic Inefficiency.

An investigation into the appropriate use of an in-principle epistemology would require, for a full treatment, a systematic investigation into what counts as a philosophical idealization. There is now a considerable body of literature in the philosophy of science on the criteria for what counts as a legitimate scientific idealization, but there is very little in the philosophical literature on what sorts of idealizations are appropriate in philosophy. Philosophy of science requires different kinds of idealization than do many other areas of philosophy. Unlike traditional epistemology, extreme skeptical objections about the reliability of evidence are off limits within the philosophy of science. No serious scientific activity can be expected to take place under conditions within which it is supposed that an evil demon could be systematically misleading us.[2] On the other hand, although there are parts of contemporary mathematical economics that are remote from economic reality, investigations into the behavior of ideally rational agents can be useful when the idealizations are a reasonable approximation to actual agents.

This tension between highly idealized scenarios and fully detailed actual examples raises questions about what should be the appropriate training for philosophers of science. Even if one acknowledges that investigating epistemological and ontological questions requires bringing to bear evidence from factual investigations, this is not by itself a reason to teach more concrete material in philosophy courses and seminars, if only because there is a strong relationship between the capacity for abstract thinking and the ability to do good philosophy of any kind.

I.2. Synopsis

Chapter 1 was originally presented at a Philosophy of Science Association symposium in 1990. It is, as far as I am aware, the first English language discussion of computer simulation methods in science by a philosopher, as contrasted with discussions about artificial intelligence. Although there is now a very large literature on the subject, the article still has some interest

---

[2] A brain in a vat could do science, even solipsistic science, but it is usually assumed in such scenarios that the brain's evidential inputs mimic those of standard science.

for its argument that scientific progress in many areas is tightly connected to progress in the application of tractable mathematics. This is one area in which detailed descriptive studies can be useful. Chapter 2 is a treatment of some of the broader intellectual consequences of computer simulations. It addresses some different senses in which the invention of computer simulations might be considered to count as a scientific revolution and discusses some ways in which science and technology interact. Chapter 3 was written as a response to claims that there is nothing philosophically new that is presented by the advent of simulation methods in science. In focusing on the principal methodological issues involved, this article is a useful corrective to some common misunderstandings about the differences between models and simulations in science. It also offers a pair of definitions for epistemic opacity, a concept that is central to many areas of computational science. Chapter 4 is the most technical of this group. Somewhat to my surprise, it has been cited many times. One reason perhaps is that it plays into the debates about the relation between simulations and experiments but in a somewhat perverse way—despite an explicit claim in the article that numerical experiments are different in important ways from material experiments, it has been cited as supporting the view that simulations are like material experiments. A final theme is that applied mathematics is not just pure mathematics applied, and that an improved understanding of how science is applied can be gained through examining these methods rather than through the philosophy of language.

Chapter 5 is the most widely cited paper in the collection. This article, which presents a way to escape the exclusion argument, became part of a longer-term program arguing the merits of diachronic approaches to emergence and exploring the possibilities of a nonatomistic ontology. The results of that program can be found in my *Emergence: A Philosophical Account* (Oxford University Press, 2016). Chapter 6, despite what I now think of as an excessively polemical tone, appears to have resonated with those who find the ambitions of metaphysical dependency relations to far exceed their usefulness. It also contains a set of criteria for what counts as emergent. Chapter 7 is an extended discussion of Mark Bedau's concept of weak emergence. It also contains an argument that this approach is, in a nontrivial way, essentially historical and cannot be accommodated within synchronic approaches to emergence. Chapter 8, which overlaps to some extent with chapter 7, lays out a taxonomy of types of emergence. It also contains the beginning of later investigations into conceptual emergence. There are other taxonomies, each with its own virtues but I have repeatedly found the one described in chapter 8 to be useful. The sixfold nature of the taxonomy reminds us of a moral that cannot be repeated often enough: there are multiple different types of emergence and no single account covers them all.

Chapters 9 and 10 form a pair centered around a result that has come to be known as "Humphreys' Paradox." The first article presents the paradox and explores a number of possible responses to it, while the second article demonstrates that the argument leads to a formal rather than an informal paradox, discusses in detail some published solutions to the paradox, and shows that none of them are successful. Although most readers have taken the arguments as a challenge to find a way out of the paradox, the principal point of the original article was to explore whether there are examples of propensities that are not correctly represented by standard probability theory and that propensities require their own theory. This theme is revisited in chapter 11, which appeared in a festschrift for a well-known statistician, and as such is probably unknown to most philosophers. The main thrust of this article is that, Quinean arguments notwithstanding, probability theory is a purely formal mathematical theory with empirical content injected via specific models that are separate from the theory.

Chapter 12 is included, although it is by now quite old, in part because it is an example of how an insight rather than pages of detailed argument can provide a solution to a philosophical problem. More important is that the position outlined here serves as a sharp contrast to those who hold that probabilistic explanation consists in the citation of probabilities and that there are no explanatory differences between contributing and counteracting causes. I have not published on explanation for many years, because I have come to believe that the fundamental philosophical concept is understanding rather than explanation. The former concept is far too grand for me to address here but chapter 13 provides a classification of explanatory schemes tuned to different modes of understanding. It also formulates a still open problem about causation regarding the point at which a causal factor can justifiably be said to have been identified. Chapter 14 is the most recent essay in the collection. It provides what I hope is a reasonably temperate and impersonal set of arguments against one way of pursuing metaphysics. It also makes a case for a certain kind of metaphysics as having a place in the philosophy of science, a view that runs counter to an antimetaphysical tradition in the area. A third theme fits with the quest to reconcile a scientifically informed realism with an equally scientifically informed empiricism. Chapter 15 is the only previously unpublished article in the collection. It describes the results from an agent-based model that uses a deformable, rather than a rigid, fitness landscape to explore how constrained maximization outperforms traditional unconstrained maximization in identifiable regions of endogenous uncertainty. One result among many is that the tragedy of the commons becomes just a special case of a much broader scenario to which specifiable solutions can be provided. I have included this paper because it shows how powerful but conceptually transparent simulations can illuminate issues in social philosophy and the philosophy of economics. More specifically, that when the assumptions underlying utility maximization fail, constraints on individualism can be beneficial. My *Extending Ourselves* book unfortunately gave short shrift to agent-based models and this article illustrates some novel features of such simulations.

Over the years, I have had the great good fortune of working with some remarkable people. There are too many to mention them all but three were of particular importance. Patrick Suppes, Wesley Salmon, and David Freedman, in very different ways, exemplified how deep ideas can be conveyed to the reader with great clarity. The essays reprinted here do not come close to their standards, but they would have been far worse without their example.

{ PART I }

Computational Science

{1}

Computer Simulations

1.1. Introduction

A great deal of attention has been paid by philosophers to the use of computers in the modeling of human cognitive capacities and in the construction of intelligent artifacts. This emphasis has tended to obscure the fact that most of the high-level computing power in science is deployed in what appears to be a much less exciting activity: solving equations. This apparently mundane set of applications reflects the historical origins of modern computing, in the sense that most of the early computers in Britain and the United States were devices built to numerically attack mathematical problems that were hard, if not impossible, to solve nonnumerically, especially in the areas of ballistics and fluid dynamics. The latter area was especially important for the development of atomic weapons at Los Alamos, and it is still true that a large portion of the supercomputing capacity of the United States is concentrated at weapons development laboratories such as Los Alamos and Lawrence Livermore.

Computer simulations now play a central role in the development of many physical sciences. In astronomy, in physics, in quantum chemistry, in meteorology, in geophysics, in oceanography, in crash analysis of automobiles, in the design of computer chips, in the planning of the next generation of supercomputers, in the discovery of synthetic pharmaceutical drugs, and in many other areas, simulations have become a standard part of scientific practice. My aim in the present paper is simply to provide a general picture of what computer simulations are, to explain why they have become an essential part of contemporary scientific methodology, and to argue that their use requires a new conception of the relation between theoretical models and their applications.[1]

---

[1] When examining this activity, we must be wary of one thing, which is that the field of computer simulation methods is relatively new and as such is rapidly evolving. Techniques that are widely used now may well be of minor interest twenty years hence, as developments in computer architecture,

Why should philosophers of science be interested in this new tool? Mostly, I think, because the way that simulations are developed and implemented forces us to re-examine a lot of what we tend to take as the right way to characterize parts of mathematically oriented methodology and theorizing. Where this re-examination takes us will become clear as we go along, but before I discuss computer simulations specifically, I want to make some general points about the role of mathematical models in physical science. Let's begin with a claim that ought to be uncontroversial, but is not given enough emphasis in philosophy of science. The claim is: *One of the primary features that drives scientific progress is the development of tractable mathematics.* Whenever you have a sudden increase in usable mathematics, there will be a concomitant sudden increase in scientific progress in the area affected. This should not really need to be pointed out, but so much emphasis is placed on conceptual changes in science that powerful instrumental changes tend to be downplayed. This kind of sudden increase in mathematical power happened with the invention of the differential and integral calculus in the middle of the seventeenth century; it happened with the sudden explosion of statistical methods at the end of the nineteenth century, and I claim that the ability to implement numerical methods on computers is, in the late twentieth century, as significant a development as those earlier inventions. But what kind of development is it? Has it introduced a distinctively different kind of method into science, as Rohrlich (1991), for example, claims, or is it simply a technologically enhanced extension of methods that have long existed? If computer simulation methods are simply numerical methods, but greatly broadened in scope by fast digital computation devices with large memory capacity, then the second "just much more of the same" view would be correct, and the situation would be similar to that in mathematics, where the introduction of computer-assisted proofs, such as were used to execute the massive combinatorial drudgery involved in the proof of the four-color theorem, is often regarded as not having changed the fundamental conception of what counts as a proof. My own view is that the situation is more complex than this simple dichotomy represents, because the introduction of computer simulation methods is not a single innovation but a multifaceted development. Let's begin with a couple of simple examples to show why mathematical intractability is an important constraint on scientific models.

---

numerical methods, and software routines take place. The specific details of different kinds of simulation methods, such as finite-difference methods and Monte Carlo methods will be explored in a future paper, and some examples of currently used simulations are given in the following paper by Rohrlich (Rohrlich 1991).

1.2. Practical and Theoretical Unsolvability of Models

Take what is arguably the most famous law of all, Newton's Second Law. This can be stated in a variety of ways, but its standard characterization is that of a second-order ordinary differential equation:

$$F = md^2y / dt^2 \qquad (1)$$

To employ this we need to specify a particular force function. In the first instance, take

$$F = GMm / R^2 \qquad (2)$$

as the gravitational force acting on a body near the earth's surface (M is the mass of the earth, R its radius). Then

$$GMm / R^2 = md^2y / dt^2 \qquad (3)$$

is easily solved. But the idealizations that underlie this simple mathematical model make it hopelessly unrealistic. So let's make it a little more realistic by representing the gravitational force as $GMm / (R + y)^2$, where y is the distance of the body from the earth's surface, and by introducing a velocity-dependent drag force due to air resistance. We obtain

$$GMm / (R + y)^2 - c\rho s (dy / dt)^2 = md^2y / dt^2 \qquad (4)$$

Suppose we want to make a prediction of the position of this body at a given time, supposing zero initial velocity and initial position $y = y_0$. To get that prediction you have to solve (4). But (4) has no known analytic solution— the move from (3) to (4) has converted a second-order, linear, homogeneous ODE into a second-order, nonlinear, homogeneous ODE, and the move from linearity to nonlinearity turns simple mathematics into intractable mathematics. Exactly similar problems arise in quantum mechanics from the use of Schrödinger's equation, where different specifications for the Hamiltonian in the schema $H\Psi = E\Psi$ lead to wide variations in the degree of solvability of the equation. For example, the calculations needed to make quantum mechanical, rather than classical, predictions in chemistry about even very simple reactions, such as the formation of hydrogen molecules when spin and vibration variables are included, are extremely difficult and have only recently been carried out. (An explicit discussion of the differences between ab initio and semiempirical methods in quantum chemistry is given below.)

You might say that this feature of unsolvability is a merely practical matter, and that as philosophers we should be concerned with what is possible in principle, not with what can be done in practice. But recent investigations into decision problems for differential equations have demonstrated that for many algebraic differential equations (ADEs) (i.e., those of the form

$$P\left(x, y_1, \ldots, y_n, y_1^{(1)}, \ldots, y_m^{(1)}, \ldots, y_1^{(n)}, \ldots, y_m^{(n)}\right) = 0$$

where P is a polynomial in all its variables with rational coefficients) it is un-decidable whether they have solutions. For example, Jaśkowski (1954) showed that there is no algorithm for determining whether a system of ADEs in several dependent variables has a solution in [0,1]. Denef and Lipshitz (1984) show that it is undecidable whether there exist analytic solutions for such ADEs in several dependent variables around a local value of x. (Further results along these lines, with references, can be found in Denef and Lipshitz 1989). Obviously, we cannot take decidability as a necessary condition for a theory to count as scientifically useful, otherwise we would lose most of our useful fragments of mathematics, but these results do show that there are in-principle, as well as practical, restrictions on what we can know to be solvable in physical theories.[2]

There is a methodological point here that needs emphasis. While much of philosophy of science is concerned with what can be done in principle, for the issue of scientific progress what is important is what can be done in practice at any given stage of scientific development. That is, because scientific progress involves a temporally ordered sequence of stages, one of the things that influences that progress is that what is possible in practice at one stage was not possible in practice at an earlier stage. If one focuses on what is possible in principle (i.e., possible in principle according to some absolute standard, rather than relative to constraints that are themselves temporally dependent), this difference cannot be represented, because the possibility-in-principle exists at both stages of development. So although what is computable in principle is important for, say, the issue of whether computational theories of the mind are too limited a representation of mental processes, what is computable in practice is the principal feature of interest for the methodologies we are considering here.

This inability to obtain specific predictions from mathematical models is a very common phenomenon, because most nonlinear ODEs and almost all PDEs have no known analytic solution. In population biology, for example, consider the Lotka-Volterra equations (first formulated in 1925)

$$dx / dt = ax + bxy$$

$$dy / dt = cy + dxy$$

---

[2] A further source of difficulty, at least in classical mechanics, involves the imposition of nonholomorphic constraints (i.e. constraints on the motion that cannot be represented in the form $f(r_1, \ldots, r_n, t) = 0$ where $\{r_i\}$ are the spatial coordinates of the particles comprising the system). For a discussion of these constraints, see Goldstein (1980), pp.11–14.

where x = population of prey, y = population of predators, a(> 0) is the difference between natural birth and death rates for the prey, $b(<0)$, $d(>0)$ are constants related to chance encounters between prey and predator, $c(<0)$ gives the natural decline in predators when no prey are available. With initial conditions $x(0) = e$, $y(0) = f$, there is no known analytic solution to the equation set.

These examples could be multiplied indefinitely, but I hope the point is clear: clean, abstract, presentations of theoretical schemas disguise the fact that the vast majority of those schemas are practically inapplicable in any direct way to even quite simple physical systems. This is not the point that models are never applicable to real systems: the point here is that even with radical idealizations, the problem of intractability is often inescapable, i.e., in order to arrive at an analytically treatable model of the system, the idealizations required would often destroy the structural features that make the model a model of that system type. This problem is widespread, and cuts across both sciences and subfields of those sciences, although it is more prevalent in some fields than in others.

These problems put severe limits on the applicability in practice of the standard, syntactically formulated method of hypothetico-deductivism, for most of the equations that represent the fundamental or derived theories of physics, chemistry, and so on cannot be used in practice to make precise deductive predictions from those representations together with the appropriate initial or boundary conditions. I should say here that I want to remain neutral as far as possible about the relative merits of the syntactic and semantic (or structuralist) reconstructions of theories. Although the semantic approach has definite advantages, both accounts are logical reconstructions of scientific practice. Because we are concerned here to stay as close as possible to considerations that present immediate problems to actual scientific practice, the debate over the merits of these reconstructions has only an indirect relevance to our interests. It is worth noting, however, that the issue of practical unsolvability means that the formulation of a theoretical model in some specific mathematical representation, rather than as a set of metamathematical structures, is an inescapable concern, and that whereas the semantic approach generally considers different linguistic formulations as mere linguistic variants of an underlying common structure, linguistic reformulations frequently have a direct impact on the ease of solvability of a mathematical representation, and hence this level cannot be ignored completely. In particular, I want to urge that what is of primary interest here is the mathematical form of equation types and not their logical form. To be specific: one could reformulate (1), (3), and (4) in a standard logical language by using variable-binding operators, thus forcing it into the standard quantified conditional form that serves as the representation of laws in the traditional syntactic approaches, but to do this would be to distort what is crucial to issues of solvability, which is the original mathematical form.

It is this predominance of mathematically intractable models that is the primary reason why computational physics (and similar methods in other sciences), which provides a practical means of implementing nonanalytic methods, constitutes a significant and, I think, a permanent addition to the mathematical methodology of science.

1.3. Definitions of Computer Simulation

Here, taken more or less at random, are some suggestions that have been made for characterizing computer simulations:

(1) "Simulation is the technique by which understanding the behaviour of a physical system is obtained by making measurements or observations of the behaviour of a model representing that system" (Ord-Smith 1975, 3).

(2) "This is what simulation is all about, i.e., experimenting with models" (Ord-Smith 1975, 3).

(3) "A precise definition of simulation is difficult to obtain . . . the term simulation will be used to describe the process of formulating a suitable mathematical model of a system, the development of a computer program to solve the equations of the model and operation of the computer to determine values for system variables" (Bennet 1974, 2).

(4) "The mathematical/logical models which are not easily amenable to conventional analytic or numeric solutions form a subset of models generally known as simulation models. A given problem defined by a mathematical/logical model can have a feasible solution, satisfactory solution, optimum solution or no solution at all. Computer modelling and simulation studies are primarily directed towards finding satisfactory solutions to practical problems" (Neelamkavil 1987, 1).

(5) "Simulation is a tool that is used to study the behaviour of complex systems which are mathematically intractable" (Reddy 1987, 162).

Because of the variety of uses to which the term "simulation" has been put, I am reluctant to try to formulate a general definition. It would be more profitable at this stage to simply explore the methods that are used under categories (1), (2), and (3) above. We can, however, formulate a working definition based on the last definition, which needs to be modified in three ways. First, simulation is a set of techniques, rather than a single tool. As the other quotations indicate, it would be hard to make a case for the view that there is an underlying unity to the set, at least at the present state of development of the field. Second, the systems that are the subject of simulations need not be complex either in structure or behavior. As we also saw earlier, mathematical intractability can affect differential or integral equations having a quite simple mathematical structure, as in the case of the motion of the body falling under the influence of gravity, subject to a velocity-dependent drag force. The behavior of this system is not unduly complex, merely hard to predict quantitatively without numerical techniques. Third, many computer simulations turn analytically intractable problems into ones that are computationally tractable, and we do not want to exclude numerical methods as a part of mathematics.

We thus arrive at the following working definition which captures what is common to almost all the simulations with which I am familiar.

Working Definition. A computer simulation is any computer-implemented method for exploring the properties of mathematical models where analytic methods are unavailable.

Some further remarks may be helpful. Although the everyday use of the term "simulation" has connotations of deception, so that a simulation has elements of falsity, this has to be taken in a particular way for computer simulations. Inasmuch as the simulation has abstracted from the material content of the system being simulated, has employed various simplifications in the model, and uses only the mathematical form, it obviously and trivially differs from the "real thing," but in this respect, there is no difference between simulations and any other kind of mathematical model, and it is primarily when computer simulations are used in place of empirical experiments that this element of falsity is important. But if the underlying mathematical model can be realistically construed (i.e., it is not a mere heuristic device) and is well confirmed, then the simulation will be as "realistic" as any theoretical representation is. Of course, approximations and idealizations are often used in the simulation that are additional to those used in the underlying model, but this is a difference in degree rather than in kind.

Next, in order for something to be a computer simulation, the whole process between data input and output must be run on a computer, whereas computational physics can involve only some stages in that process, with the others being done "by hand." Third, because computer simulations are usually oriented toward approximate solutions rather than exact solutions, they can be viewed as optimization devices that sometimes involve satisficing criteria. This approach underlies the variational method mentioned later, it underlies the simulated annealing method frequently used in connectionist models of perception and problem solution (see McClelland and Rumelhart 1986, especially chap. 6), and it underlies many other intuitive "good enough" criteria used in other areas.

1.4. Can Computer Simulations Be Identified
with Numerical Methods?

What is computer simulation? The terminology is so widely used that it is hard to find a core meaning, but here are some central uses:

(1) To provide solution methods for mathematical models where analytical methods are presently unavailable.
(2) To provide numerical experiments in situations where natural experimentation is inappropriate (for practical reasons) or unattainable (for physical reasons). Under the former lie experiments that are too costly, too uncertain in their outcome, or too time consuming. Under the latter lie such experiments as the rotation of angle of sight of galaxies, the formation of thin disks around black holes, and so forth.
(3) To generate and explore theoretical models of natural phenomena.

It may seem that use (1) is simply the use of numerical methods for solution purposes. To examine this claim, we need some definitions. *Numerical mathematics* is concerned with obtaining numerical values of the solutions to a given mathematical problem. *Numerical methods* is the part of numerical mathematics concerned with finding an approximate, feasible solution. *Numerical analysis* has as its principal task the theoretical analysis of numerical methods and the computed solutions, with particular emphasis on the error between the computed solution and the exact solution.

Can we identify numerical methods with computer simulations? Not directly, because there are at least two additional features that a numerical method must have if it is to count as a computer simulation. First, the numerical method must be applied to a specific scientific problem in order to be part of a computational simulation. Second, the method must be computable in real time and be actually implemented on a concrete machine.

Beyond this, there is an important potential distinction between uses (1) and (3). In (1), the development of the model is made along traditional lines: some more or less fundamental theory is brought to bear on the phenomenon, theory which at least in its abstract, general form is well understood and confirmed.

Deductive consequences are drawn out from this theory to bring the general theory into contact with the specific area under investigation, and then the computational implementation of these consequences constitutes the simulation of the system. In contrast, in use (3), the development of the models is partly empirical, partly theoretical, and partly heuristical, with exploration and feedback from the simulation playing an important role in this development.

This distinction is not clear-cut, and especially in use (3), elements from uses (1) and (2) often play a significant role. The difference is similar to a distinction that is often drawn in quantum chemistry between ab initio methods and semiempirical methods (see, e.g., McWeeny and Sutcliffe 1969, chap. 9) Three kinds of treatments can be used to predict the energy levels of molecular orbitals. Ab initio methods use the actual Hamiltonian for the system in Schrödinger's equation. Idealizations are made, such as a fixed nucleus, only electrostatic interactions between particles, and nonrelativistic calculations, and these idealizations are often drastic, but the goal is to represent as many of the important features of the molecules as possible. Then, using a "trial function," the equation calculates the solution "exactly." Semiempirical methods estimate some parameters in the orbital states that are difficult to calculate directly by empirical data or by numerical approximation, and then proceed as in the ab initio case. Model-level methods use a Hamiltonian that deliberately omits some important influences on the energy levels, such as interelectron interactions.

The distinction here between ab initio methods and model-level methods seems to me to be quite arbitrary, since both use idealizations, and the interesting difference is that between ab initio methods and semiempirical methods, and this is an appropriate place to discuss the differences between fundamental and phenomenological models. This distinction reflects the "bottom up" and "top down" methods familiar from other areas of methodology, and there is a significant divergence of views about whether models should be constructed on the basis of some underlying general theoretical considerations, or whether instrumentally successful but theoretically ungrounded models should be used when the theoretical approach is infeasible. Both kinds are used in simulations and I see no reason to deny the appropriateness of either. I choose to focus on fundamental models here, primarily for two reasons. The first is one of expertise, or lack of it. Phenomenological models are usually highly specific devices constructed for the purpose of representing some specific phenomenon. A great deal of physical, chemical, or biological knowledge goes into their assessment, justification, and use (this is one area where "physical intuition" is clearly an important consideration) and for this reason, such simulations can be assessed only by those actively working with them. The second reason for emphasizing fundamental models here is that this makes a comparison of simulation methodology with traditional philosophical views on theory structure and application much easier, for the latter is oriented almost exclusively toward fundamental theory. This said, a few remarks about the relation between the two approaches in the case of ab initio and semiempirical models might be appropriate. (Semiempirical models are not the same as phenomenological models, in that the former are still guided to a considerable extent by theory, but for the first reason just mentioned, I am not in a position to address phenomenological models in any detail.)

One important result of the availability of large-scale computational power is that whereas the use of semiempirical methods, or many idealizations in models, were once forced upon chemists because the model had to result in tractable analytic mathematics, the idealizations made in ab initio methods now need not be determined primarily by that constraint, but are set by (a) limits on computational power available, (b) the ability to mathematically represent in the Hamiltonian complex influences on the energy levels, (c) the availability of numerical methods to approximate the representations in (b). This is a clear example of computational chemistry: the use of computers to allow one to treat models that could not be used without them. Indeed, these methods illustrate an interesting trade-off: These numerical methods allow one to deal with more realistic theories, and the increased use of approximations in the mathematics allows a decreased use of idealizations in the physics. This still leaves the treatment of most molecules currently outside the scope of ab initio methods, and given the restrictions due to (a) that are discussed below, no purely ab initio method will ever be fully computationally feasible, but the important point is that more and more systems that were once untreatable by fundamental approaches can now have theoretically justified quantum mechanical methods brought to bear on them.

Compare this with methods that rely on the variation theorem. (See Eyring, Walter, and Kimball 1944 for a development of this theorem.) The theorem states, "If a normalized trial function S satisfies the relevant boundary conditions but is otherwise arbitrary, then $<S/H/S> \geq E_0$, where E_0 is the lowest eigenvalue, the equality applying when S is an exact solution." Then the best wave function is obtained by varying the parameters in a trial function until the lowest energy is obtained. Here, the computational methods allow exploratory investigations that would not be possible without computers, and these are different from theoretically based methods in that although theory may be used as a guide to which parametric family of functions to explore (Gaussian or Slater orbitals are usually used, however, making the contribution of theory minimal), the final result is a matter of computational trial and error rather than explicit theoretical derivation. Moreover, "A wave function that gives a good [estimate of the] energy does not necessarily give a particularly good value for another quantity, for example the dipole moment, whose expectation value may arise principally from somewhat different regions of space." (McWeeny and Sutcliffe 1969, 235). I also find these figures from McWeeny and Sutcliffe (1969, 239) revealing: abstract developments of quantum mechanics require an infinite set of basis vectors to represent states. For the finite basis sets that actual applications need, suppose that m atomic orbitals are used (in the linear combination of atomic orbitals representation of molecular orbitals—the LCAO method). Then one needs $p = m(m+1)/2$ distinct integrals to calculate one-electron Hamiltonians, and $q = p(p+1)/2$ distinct integrals to calculate electron interaction terms. This gives

$$
\begin{array}{ccccc}
m = & 4 & 10 & 20 & 40 \\
q = & 55 & 1540 & 22155 & 336610
\end{array}
$$

This is a clear case where computational constraints, which are extratheoretical and here involve primarily memory capacity, place severe limitations on what can be done at any given stage of technological development. This is different in principle, I think, from the constraints that the older analytic methods put on model development, because there new mathematical techniques had to be developed to allow more complex models, whereas in many cases in computational science, the mathematics stays the same, and it is technology that has to develop. The use of trial orbitals that I mentioned earlier in connection with the variational method seems to show that a very crude model can give an apparently realistic representation of the system. That is, in deciding upon the appropriate potential energy function to use in the Hamiltonian, suppose we choose one corresponding to a Slater atomic orbital of the form

$$V(r) = -fn/r + [n(n-1) - l(l+1)] / 2r^2$$

where f is a parameter representing the effective field affecting the electron. Then using these atomic orbitals as the finite basis, we have to decide where the expansion of the state function will be truncated. Then, given various trials R, and trial orbitals S for another electron, we have to minimize the energy

$$E = 2 < R/h/R > + (2 < RS/g/RS > - < RS/g/SR >)$$

Although there is a great deal of computation involved here, and certainly trial and error "experimentation," there is also a good deal of theory lying in the background to justify the method, and even though the atomic orbitals used are pretty crude approximations, they are still guided by a physical model that has some theoretical justification.

I am thus going to treat each of uses (1), (2) and (3) above as part of computational physics (chemistry, etc.), and to consider computer simulation as a subset of the methods of computational science. Much more needs to be said about what is special to simulations, but I hope that the example just discussed shows that the interplay between theory, experiment, and computation in computational science entails that it is not to be identified with numerical methods, and a fortiori, neither should computer simulations.[3]

References

Bennett, A. Wayne. 1974. *Introduction to Computer Simulation*. St. Paul, MN: West.

Denef, J., and L. Lipshitz. 1984. "Power Series Solutions of Algebraic Differential Equations." *Mathematische Annalen* 267: 213–38.

---

[3] Research for this paper was supported by NSF grant DIR-8911393. I should like to thank Fritz Rohrlich for helpful discussions in connection with the PSA symposium.

————. 1989. "Decision Problems for Differential Equations." *Journal of Symbolic Logic* 54: 941–50.

Eyring, Henry, John Walter, and George E. Kimball. 1944. *Quantum Chemistry*. New York: J. Wiley and Sons.

Goldstein, Herbert. 1980. *Classical Mechanics*. 2nd ed. Reading, MA: Addison-Wesley.

Jaśkowski, Stanisław. 1954. "Example of a Class of Systems of Ordinary Differential Equations Having No Decision Method for Existence Problems." *Bulletin de l' Academie Polonaise des Sciences*, Classe III, 2: 155–57.

McClelland, James L., and David E. Rumelhart. 1986. *Parallel Distributed Processing*. 2 vols. Cambridge: MIT Press.

McWeeny, R., and B. T. Sutcliffe. 1969. *Methods of Molecular Quantum Mechanics*. New York: Academic Press.

Neelamkavil, Francis. 1987. *Computer Simulation and Modelling*. New York: J. Wiley.

Ord-Smith, R. J., and J. Stephenson. 1975. *Computer Simulation of Continuous Systems*. New York: Cambridge University Press.

Reddy, R. 1987. "Epistemology of Knowledge-Based Systems." *Simulation* 48: 161–70.

Rohrlich, Fritz. 1991. "Computer Simulation in the Physical Sciences." In *PSA: Proceedings of the Biennial Meeting of the Philosophy of Science Association 1990*, vol. 2, edited by Arthur Fine, Mickey Forbes, and Linda Wessels, 497–506. East Lansing, Michigan: Philosophy of Science Association.

{ 2 }

Computational Science and Its Effects

2.1. Introduction

The rise of computational science, which can be dated, somewhat arbitrarily, as beginning around 1945–1946,[1] has had effects in at least three connected domains—the scientific, the philosophical, and the sociotechnological context within which science is conducted.[2] Some of these effects are secondary, in the sense that disciplines such as complexity theory would have remained small theoretical curiosities without access to serious computational resources. Other effects, such as the possibility of completely automated sciences, are longer term and will take decades to alter the intellectual landscape. I shall provide here some examples of fine-grained philosophical effects as well as examples of more sweeping social and intellectual consequences that will suggest both the different ways of thinking that these methods require and a hint at how far-reaching they are.

First, we need a framework. In their paper "Complex Systems, Modelling and Simulation," Sylvain Schweber and Matthias Wächter 2000 suggested that the introduction and widespread use of computational science constitutes what they call a "Hacking revolution" in science and that Hacking's use of "styles of reasoning," a concept which originated with the historian of science A. C. Crombie, can give us useful insights into these methods. Schweber and Wächter have many useful things to say about simulations and related

---

[1] I identify its origins with the use of electronic computers to perform Monte Carlo calculations at Los Alamos and John Mauchley's suggestion that ENIAC could be used for difference equation simulations, rather than for just routine arithmetical calculations. See Metropolis 1993, 127 for the second point. I do not vouch for the accuracy of Metropolis's recollections on this point although the exact historical turning point, if indeed "exact" ever makes sense in historical claims, is unimportant. For those interested in technoscience, I note that the innovation had its origins at Los Alamos and other military research institutions rather than in industrial applications.

[2] There are other domains it has affected, but I shall restrict my discussion to these three.

methods, but Hacking's framework does not sit well on computational science. Let me say why.

Hacking revolutions have four principal characteristics: First, "they transform a wide range of scientific practices and they are multi-disciplinary." In this, they are different from the more familiar Kuhnian revolutions or the shifts in theoretical research programs suggested by Imre Lakatos, which tend to be limited to single scientific disciplines. Computational science satisfies this first condition, most notably because its methods are largely transdisciplinary. Second, a Hacking revolution leads to new institutions designed to foster the new practices. The Santa Fe Institute is an example of this second feature.[3] The third characteristic is that "the revolution is linked with substantial social change." Changes in the social structure of science are hard to separate from more general societal changes introduced by computers, but it is true that the social structure of science has been affected by the easy electronic exchange of ideas, the dominant role of programmers in a research group, and remote access to supercomputers. The fourth characteristic is that "there can be no complete, all-encompassing history of such revolutions."

Although there is merit in the concept of a Hacking revolution, I shall not use it here for two reasons. The first is that Hacking revolutions share their second and third components with Kuhnian revolutions (because of the tight link between the intellectual and sociological aspects of Kuhn's position, these components are satisfied almost by default in a Kuhnian revolution). And the fourth condition is almost trivially true of any such historical episode. This leaves only the multidisciplinary aspect, which is important but lacks fine structure. Second, let me make a distinction between *replacement revolutions* and *emplacement revolutions*. Replacement revolutions are the familiar kind in which an established way of doing science is overthrown and a different set of methods takes over. Emplacement revolutions occur when a new way of doing science is introduced which largely leaves in place existing methods. The introduction of laboratory experimentation was an emplacement revolution in the sense that it did not lead to the demise of theory or of observation. Similarly, the rise of computational science constitutes an emplacement revolution. This is not to say that theory and experiment are not affected by computational approaches, because certain theoretical methods have now been taken over by computational methods, and many experiments are now computer assisted, but theory and experiment have not been abandoned and considered scientifically unacceptable in the way that the replacement revolutions of Copernicus over Ptolemy, Newton over Descartes, or Darwin over gradualism resulted in the untenability of the previous approaches.

---

[3] Although the Institute has recently announced that because complexity science is now well established, it must move in new directions.

What of styles of reasoning? Here are six cases, originally identified by Crombie, that are cited by Hacking, as examples of the genre:

(a) The simple method of postulation exemplified by the Greek mathematical sciences
(b) The deployment of experiment both to control postulation and to explore by observation and measurement
(c) Hypothetical construction of analogical models
(d) Ordering of variety by comparison and taxonomy
(e) Statistical analysis of regularities of populations, and the calculus of probabilities
(f) The historical derivation of genetic development (Hacking 1992, 4)

Hacking then says: "Every style of reasoning introduces a great many novelties including new types of objects; evidence; sentences, new ways of being a candidate for truth or falsehood; laws, or at any rate modalities; possibilities. One will also notice, on occasion, new types of classification, and new types of explanations. . . . Hence we are in a position to propose a necessary condition for being a style of reasoning: each style should introduce novelties of most or all of the listed types and should do so in an open-textured, ongoing, and creative way" (Hacking 1992, 11–12, slightly reformatted). One could squeeze computational science into this framework because three of the five criteria are satisfied—novelties of evidence, sentences, and possibilities—but as I shall argue, laws are the wrong vehicle for understanding what is distinctive about computational science, and the novel objects are better understood as novel representations. Moreover, "style of reasoning" has an anthropocentric flavor that is best avoided in this context. So "style of reasoning" is not a good fit. Instead, I shall use the term "technique" in what follows.

2.2. The Main Issue

Let me put the principal philosophical novelty of these methods in the starkest possible way: computational science introduces new issues into the philosophy of science because it uses methods that push humans away from the center of the epistemological enterprise. In doing this, it is continuing a historical development that began with the use of clocks and compasses, as well as the optical telescope and microscope, but it is distinctively different in that it divorces reasoning, rather than perceptual, tasks from human cognitive capacities. There were historical ancestors of computational science, such as astrolabes and orreries, but their operation was essentially dependent upon human calculations.

Until recently, science has always been an activity that humans carry out and analyze. It is also humans that possess and use the knowledge produced

by science. In this, the philosophy of science has followed traditional epistemology, which, with a few exceptions such as the investigation of divine omniscience, has been the study of human knowledge. Locke's *Essay Concerning
Human Understanding*, Berkeley's *A Treatise Concerning the Principles of
Human Knowledge*, Hume's *A Treatise of Human Knowledge*, Reid's *Essays on
the Intellectual Powers of Man* are but a few examples; the Cartesian and Kantian
traditions in their different ways are also anthropocentric.[4] In the twentieth
century, the logical component of logical empiricism broke free from the psychologism of earlier centuries, but the empiricist component prevented a complete separation.[5] Two of the great alternatives to logical empiricism, Quine's
and Kuhn's epistemologies, are rooted in communities of human scientists and
language users. Even constructive empiricism and its successor, the empirical stance, are firmly anchored in human sensory abilities (van Fraassen 1980,
2004). There are exceptions to this anthropocentric view, such as Popper 1972
and Ford, Glymour, and Hayes 2006, but the former's World 3 is too abstract
for our concerns and the latter's artificial intelligence orientation does not address the central issues of computational science.[6]

At this point I need to draw a distinction. Call the current situation within
which humans deal with science that is carried out at least in part by machines
the *hybrid scenario*, and the more extreme situation of a completely automated
science replacing the science conducted by humans the *automated scenario*.
This distinction is important because in the hybrid scenario, one cannot completely abstract from human cognitive abilities when dealing with representational and computational issues. In the automated scenario one can, and it is
for me the more interesting philosophical situation, but in the near term we
shall be in the hybrid scenario and so I shall restrict myself here to that case. It
is because we are in the hybrid scenario that computational science constitutes
an emplacement revolution. If the automated scenario comes about, we shall
then have a replacement revolution.

For an increasing number of fields in science, an exclusively anthropocentric epistemology is no longer appropriate because there now exist superior, nonhuman, epistemic authorities. So we are now faced with a problem,

---

[4] A Kantian approach can be generalized to nonhuman conceptual categories, although the extent
to which humans could understand those alien categories is then a version of one philosophical challenge faced by computational science.

[5] Carnap's *Aufbau* ([1928] 1967) allows that a physical basis could be used as the starting point
of the reconstruction procedure, but adopts personal experiences as the autopsychological basis. The
overwhelming majority of the literature in the logical empiricist tradition took the human senses as
the ultimate authority.

[6] One can usefully borrow Popper's thought experiment in which all of the world's libraries are
destroyed and ask how much of contemporary science would be affected if neutron bombs shut down
all of the world's computers. Much of "big science," especially in physics and astrophysics, would be
impossible to carry out.

which we can call the *anthropocentric predicament*, of how we, as humans, can understand and evaluate computationally based scientific methods that transcend our own abilities and operate in ways that we cannot fully understand. Once again, this predicament is not entirely new because many scientific instruments use representational intermediaries that must be tailored to human cognitive capacities. With the hybrid situation, the representational devices, which include simulations and computationally assisted instruments such as automated genome sequencing, are constructed to balance the needs of the computational tools and the human consumers. We can call the general problem of inventing effective intermediaries the *interface problem* and it is a little-remarked-upon aspect of scientific realism when we access the humanly unobservable realm using instruments. Just as scientific instruments present philosophy with one form of the metaphysical problem of scientific realism and its accompanying epistemological problems, so computational science leads to philosophical problems that are both epistemological, a feature that has been emphasized by Eric Winsberg and Johannes Lenhard,[7] and metaphysical.

2.3. What Is Metaphysically Different about Computational Science

The essence of computational science is providing computationally tractable representations; objects that I have elsewhere called computational templates.[8] It is an important feature of templates that they are transdisciplinary. The philosophical literature on scientific laws, with its emphasis on counterfactuals, nomological necessity, logical form, and so on, often does not stress the fact that the fundamental laws of a science are uniquely characteristic of that science. Although Newton's laws applied to any material object in the eighteenth century, they did not characterize biological objects qua biological objects in the way that they did characterize what it was to be a physical object. Nowadays, the Hardy-Weinberg law is a characteristic feature of population biology, and it makes no sense in chemistry or physics.[9]

I mentioned above that laws are the wrong vehicle for understanding computational science. The reason for this is connected with the point just noted that scientific laws are intimately tied to a particular science and its subject matter, whereas the emphasis of computational science is on transdisciplinary

---

[7] See, e.g., Winsberg 2001, 2003; Lenhard 2007.

[8] See Humphreys 2002; 2004, Chap. 3; 2008.

[9] To prevent misunderstanding, I note that although the term "law" is used for such things as the weak and strong laws of large numbers in probability theory, because these are purely mathematical results, this is a courtesy use of the term "law." They lack at least the nomological necessity possessed by scientific laws.

representations. There are some candidates for laws of this transdisciplinary type in complexity theory, such as Zipf's Law, a power law that reasonably accurately describes the distribution of city sizes, network connection densities, the size of forest fires, and a number of other phenomena that are the result of scale-invariant features. Just as theory and experiment involve techniques that are to a greater or lesser extent subject matter independent, so too does computational science. This cross-disciplinary orientation has at least two consequences that are worth mentioning. First, it runs counter to the widely held view that models are local representations. It is, of course, true that many models are far less general than theories, but the existence of widely used computational templates suggests that the disunity-of-science thesis that often accompanies the "models are local" thesis is simply wrong about the areas of contemporary science that lend themselves to the successful use of such templates. Second, it runs orthogonally to the traditional reductionist approach to understanding. Reduction suggests to us that we can better understand higher-level systems by showing how they can be reduced to, how they can be explained in terms of, lower-level systems. Computational templates suggest that we can gain understanding of systems without pursuing reduction by displaying the common structural features possessed by systems across different subject domains. In saying this, I am not claiming that these transdisciplinary representations did not exist prior to the introduction of computational science. What the latter development did was to allow the vastly increased use of these techniques in ways that made their application feasible.

I can illustrate the issue involved using as an example agent-based simulations. Agent-based simulations are in certain ways very different from what one might call equation-based simulations. It is a common, although not universal, feature of agent-based models that emergent macro-level features appear as a result of running the simulation, that these features would not appear without running the simulation, that new macro-level descriptions must be introduced to capture these features, and that the details of the process between the model and its output are inaccessible to human scientists. No traditional modeling methods address the first, second, and fourth features of these simulations. Let me elaborate a little on how the third point plays out in this context. The situation has been nicely captured by Stephen Weinberg: "After all, even if you knew everything about water molecules and you had a computer good enough to follow how every molecule in a glass of water moved in space, all you would have would be a mountain of computer tape. How in that mountain of computer tape would you ever recognize the properties that interest you about the water, properties like vorticity, turbulence, entropy, and temperature?" (Weinberg 1987, 434). Many of the "higher level" conceptual representations needed to capture the emergence of higher-level patterns do already exist in other theoretical representations; they are the starting point for what Ernest Nagel called inhomogeneous reductions. With other agent-based models the situation is different because the simulation itself will, in some cases, construct a novel macro-level feature. It is this constructivist aspect of simulations, one that runs in the opposite direction to the traditional reductionist tendency of theories, that is a characteristic feature of agent-based models in particular, although it also can be a focus of equation-based models. Constructivism was memorably described in Anderson 1972 and is a key element of the arguments presented in Laughlin and Pines 2000.[10] These emergent patterns in computer simulations form the basis for what Mark Bedau has characterized as "weak emergence" (Bedau 1997) and traditional human modeling techniques will not generate them from the agent base. They can only be arrived at by simulation.

This emphasis on higher-level patterns is not restricted to computational science or to emergence. It is a feature of multiply realizable systems and of physical systems in which universality is exhibited. As another example, Niklas Luhmann, the German sociologist, has persuasively argued for the irrelevance of individual humans in various functional systems.[11] For example, within consumer economies, it is irrelevant who purchases the pack of cigarettes—they can be male, female, Chilean or Chinese, middle-aged or old, white collar or blue collar—all that matters is that the relevant economic communications take place. Indeed, Luhmann's work is a striking example of a research program within which the importance of humans as individuals is severely diminished and the emphasis placed on the autonomy of higher-level features. Luhmann was an early advocate of autopoiesis, a process that leads to self-organizing systems. One of the core features of self-organizing systems is that there is no central organizing force controlling the system. The American Stryker forces that are currently operating in Iraq and Afghanistan are a contemporary example of the movement toward engineering systems of this kind. Every member of a squad is issued a radio or other communications device, with the result that information is no longer concentrated in and processed through a central command system, and the lowest-ranking infantryman will often be better informed of the dynamically evolving state than will the commanding officer. Because the command hierarchy is still in place, the tension between the two is understandably the subject of much debate.

Computational science can also produce significant shifts in specific sciences. For example, general equilibrium theory, which dominated neoclassical economics for decades, is now being challenged by rival approaches such as agent-based microeconomics and evolutionary game theory. These developments are sensible because humans tend to have a good insight into

---

[10] The use of generative mechanisms as an element of constructivism is noted in Küppers and Lenhard 2006.

[11] Luhmann's culminating work is Luhmann 1997, which is not yet available in an English translation. I am grateful to Tiha von Ghyczy for conversations about various aspects of Luhmann's thought.

the nature of social and economic relations between individuals and much less of a firm grip on the kind of grand hyperidealized theory that was once dominant.

2.4. What Is Epistemically New about Computational Science

The rise of computational science has allowed an enormous increase in scientific applications. But this expansion has also been accompanied by a shift in emphasis from what is possible in principle to what is possible in practice, with the countervailing result that the domain of science has also shrunk. Let me explain.

2.4.1. IN PRACTICE, NOT IN PRINCIPLE

One feature of computational science is that it forces us to make a distinction between what is applicable in practice and what is applicable only in principle. Here the shift is, first, from the complete abstraction from practical constraints that is characteristic of much of traditional philosophy of science and, second, from the kind of bounded scientific rationality that is characteristic of the work of Simon and Wimsatt (1974), within which the emphasis tends to be on accommodating the limitations of human agents. Ignoring implementation constraints can lead to inadvisable remarks. It is a philosophical fantasy to suggest, as Manfred Stöckler does, that "In principle, there is nothing in a simulation that could not be worked out without computers" (2000, 368).[12]

In saying this I am not in any way suggesting that in-principle results are not relevant in some areas. They clearly are; there are also other issues to which the philosophy of science needs to devote attention. One of the primary reasons for the rapid spread of simulations through the theoretically oriented sciences is that simulations allow theories and models to be applied in practice to a far greater variety of situations. Without access to simulation, applications are sometimes not possible; in other cases the theory can be applied only to a few stylized cases.

Within philosophy, there is a certain amount of resistance to including practical considerations, a resistance with which I can sympathize and I am by no means suggesting that the investigation of what can (or cannot) be done in principle is always inappropriate for the philosophy of science. One source of resistance to using in-practice constraints is already present in the

---

[12] The first versions of Thomas Schelling's agent-based models of segregation, and the first versions of Conway's Game of Life were done "by hand," but almost all contemporary simulations require abilities that go far beyond what is possible by the unaided human intellect.

tension between descriptive history of science and normative philosophy of science, and in the tension between naturalistic approaches (which tend to mean different things to different people) and more traditional philosophy of science. But the appeal to in-principle arguments involves a certain kind of idealization, and some idealizations are appropriate whereas others are not. A long-standing epistemological issue involves the limits of knowledge. Are there things that we cannot know, and if so, can we identify them? There surely cannot be any question that this is a genuine philosophical problem. Of course, it is not new—Kant famously gave us answers to the question. The question of what we can know, or more accurately, what we can understand, has been transformed by the rise of computational science and it is partly a question of what idealizations can legitimately be used for epistemic agents. We already have experience in what idealizations are appropriate and inappropriate for various research programs. The move away from hyperrational economic agents in microeconomics to the less idealized agents mentioned earlier is one well-known example. For certain philosophical purposes, such as demonstrating that some kinds of knowledge are impossible even in principle, in-principle arguments are fine. But just as humans cannot in principle see atoms, neither can humans in principle be given the attributes of unbounded memory and computational speed. This is the reason underlying epistemic opacity, one of the key epistemological features of the new methods.

2.4.2. EPISTEMIC OPACITY

One of the key features of computational science is the essential epistemic opacity of the computational process that leads from the abstract model underlying the simulation to its output. Here a process is epistemically opaque relative to a cognitive agent X at time t just in case X does not know at t all of the epistemically relevant elements of the process. A process is essentially epistemically opaque to X if and only if it is impossible, given the nature of X, for X to know all of the epistemically relevant elements of the process.[13] For a mathematical proof, one agent may consider a particular step in the proof to be an epistemically relevant part of the justification of the theorem, whereas to another, the step is sufficiently trivial to be eliminable. In the case of scientific instruments, it is a long-standing issue in the philosophy of science whether the user needs to know details of the processes between input

---

[13] In my 2004, I used only the straightforward "epistemically opaque" terminology. I now think that distinguishing between the weaker and stronger senses is useful. It is obviously possible to construct definitions of "partially epistemically opaque" and "fully epistemically opaque" which the reader can do himself or herself if so inclined. What constitutes an epistemically relevant element will depend upon the kind of process involved.

and output in order to know that what the instruments display accurately represents a real entity.

Within the hybrid scenario, no human can examine and justify every element of the computational processes that produce the output of a computer simulation or other artifacts of computational science. This feature is novel because, prior to the 1940s, theoretical science had not been able to automate the process from theory to applications in a way that made the details of parts of that process completely inaccessible to humans. Many, perhaps all, of the features that are special to simulations are a result of this inability of human cognitive abilities to know in detail what the computational process consists in. The computations involved in most simulations are so fast and so complex that no human or group of humans can in practice reproduce or understand the processes. Although there are parallels with the switch from an individualist epistemology, within which a single scientist or mathematician can verify a procedure or a proof, to social epistemology, within which the work has to be divided between groups of scientists or mathematicians, so that no one person understands all of the process, the sources of epistemic opacity in computational science are very different.

One of the major unresolved issues in many areas of computational science is whether the invention of new mathematical techniques might eventually replace some of these computational methods. I have frequently heard the suggestion that if we introduced a new class of functions that were solutions to the existing, currently intractable model, this would not change the way the model relates to the world. In fact it would, because with the availability of analytic solutions, the epistemic opacity of the relation between the model and the application would disappear. Moreover, even if this were to happen, the fact that the computational methods are, during our era, an unavoidable part of scientific method makes them of philosophical interest, just as the use of the Ptolemaic apparatus for computing planetary orbits is still of philosophical interest.

There are aspects of computational science that are simply not addressed by either of the two traditional philosophical accounts of theories. The traditional syntactic account of theories distinguished between some types of theories: those that were recursively axiomatizable, those whose axioms sets are only recursively enumerable, and a few other types. Computer scientists have since added to this classification, in moving from the simple issue of (Turing) computability to measures of theoretical computational complexity, such as P, NP, P-SPACE, and many others. This refinement can be incorporated within the syntactic account of theories. Other issues about the power of different computational architectures that are also relevant to computational science cannot be so incorporated. It is possible that if operational quantum or biological computers are built, a number of scientifically intractable problems will become tractable, opening up new areas of research. This is not an issue that is in any way addressed by traditional modeling techniques and although philosophical discussions of quantum computing have not been motivated much by issues in the area of simulations, the area is novel and is relevant to computational science. (See, e.g., Mermin 2007.)

2.4.3. THE LINK BETWEEN SCIENCE AND TECHNOLOGY

The final issue to be addressed is the way in which progress in various sciences is now tied to technological advances in ways that go beyond the dependencies produced by a reliance on instrumentation. Computer simulations are crucially dependent upon computational load issues, and science must often wait until the next generation of machines is developed for these load demands to be accommodated. Technological issues arise in other ways as well: there are problems of extending models when substantial chunks of existing code are written in languages that are not compatible with other modules in the software and are thus hard to integrate into later research; the former may require obsolete hardware to run. Philosophers of science are free to abstract from these issues, but then in some areas of science their accounts will simply misrepresent how progress is made.

Even with idealizations, these computational features are relevant. Here is one particular example: Determining energy levels is a core interest for molecular chemists. Physical chemistry employs quantum mechanics as its basic theoretical apparatus, but ab initio calculations of the energy levels are impossible to carry out for any but the smallest molecules. The simple valence bond and molecular orbital models do not provide accurate predictions even for hydrogen molecules, so they have to be supplemented with dozens of extra terms to account for various features. They therefore employ multiple approximations and are heavily computational. So the approximations chosen in the Hartree-Fock self-consistent field approach, a standard method of calculating ground state energies in ab initio quantum chemistry, are inextricably linked with the degree to which those calculations can actually be carried out in practice. On the other side there is now a growing sense that a different problem has arisen; that new techniques need to be developed to effectively exploit the massive computational power that is now available in many areas.[14]

Although some skepticism has been expressed about the novelty of computer simulations and related techniques (e.g., Stöckler 2000; Frigg and Reiss 2009; for a response see Humphreys 2009), there is more than enough evidence to support claims that they constitute an important addition to the techniques of science, on a par with theoretical representations and experiment. I hope

---

[14] "Rationale for a Computational Science Center," unpublished report, University of Virginia, March 2007.

that what I have said above gives some insight into the more general intellectual consequences of these new ways of doing science.

References

Anderson, Philip W. 1972. "More Is Different." *Science* 177: 393–96.

Bedau, Mark. 1997. "Weak Emergence." *Philosophical Perspectives* 11: 375–99.

Carnap, Rudolph (1928) 1967. *Der logische Aufbau der Welt*. Berlin. Translated as *The Logical Structure of the World* by Rolf George. Berkeley: University of California Press.

Ford, Kenneth, Clark Glymour, and Patrick Hayes. 2006. *Thinking about Android Epistemology*. Menlo Park, CA: AAAI Press.

Frigg, Roman, and Julian Reiss. 2009. "The Philosophy of Simulation: Hot New Issues or Same Old Stew?" *Synthese* 169: 593–613.

Hacking, Ian. 1992. "'Style' for Historians and Philosophers." *Studies in History and Philosophy of Science* 23: 1–20.

Humphreys, Paul. 2002. "Computational Models." *Philosophy of Science* 69: S1–S11.

———. 2004. *Extending Ourselves: Computational Science, Empiricism, and Scientific Method*. New York: Oxford University Press.

———. 2009. "The Philosophical Novelty of Computer Simulation Methods." *Synthese* 169: 615–26.

Küppers, Günter, and Johannes Lenhard. 2006. "From Hierarchical to Network-Like Integration: A Revolution of Modeling Style in Computer-Simulation." In *Simulation: Pragmatic Constructions of Reality*, edited by Johannes Lenhard, Günter Küppers, and Terry Shinn, 89–106. Berlin: Springer.

Laughlin, R. B., and David Pines. 2000. "The Theory of Everything." *Proceedings of the National Academy of Sciences* 97: 28–31.

Lenhard, Johannes. 2007. "Computer Simulations: The Cooperation between Experimenting and Modeling." *Philosophy of Science* 74: 176–94.

Luhmann, Niklas. 1997. *Die Gesellschaft der Gesellschaft*. Frankfurt am Main: Suhrkamp.

Mermin, N. David. 2007. *Quantum Computer Science*. Cambridge: Cambridge University Press.

Metropolis, Nicholas. 1993. "The Age of Computing: A Personal Memoir." In *A New Era in Computation*, edited by N. Metropolis and Gian-Carlo Rota, 119–30. Cambridge, MA: MIT Press.

Popper, Karl. 1972. "Epistemology without a Knowing Subject." In *Objective Knowledge: An Evolutionary Approach*, 106–52. Oxford: Oxford University Press.

Redhead, Michael. 1980. "Models in Physics." *British Journal for the Philosophy of Science* 31: 145–63.

Schweber, Sam, and Matthias Wächter. 2000. "Complex Systems, Modelling and Simulation." *Studies in History and Philosophy of Modern Physics* 31 (4): 583–609.

Stöckler, Manfred. 2000. "On Modeling and Simulations as Instruments for the Study of Complex Systems." In *Science at Century's End: Philosophical Questions on the Progress and Limits of Science*, edited by Martin Carrier, Gerald Massey, and Laura Ruetsche, 355–73. Pittsburgh: University of Pittsburgh Press.

van Fraassen, Bas. 1980. *The Scientific Image*. Oxford: Clarendon Press.

———. 2004. *The Empirical Stance*. New Haven: Yale University Press.

Weinberg, Stephen. 1987. "Newtonianism, Reductionism, and the Art of Congressional Testimony." *Nature* 330: 433–37.

Winsberg, Eric. 2001. "Simulations, Models, and Theories: Complex Physical Systems and Their Representations." *Philosophy of Science* 68: S442–S454.

———. 2003. "Simulated Experiments: Methodology for a Virtual World." *Philosophy of Science* 70: 105–25.

Wimsatt, William C. 1974. "Complexity and Organization." In *PSA 1972: Proceedings of the 1972 Biennial Meeting of the Philosophy of Science Association*, edited by Kenneth Schaffner and Robert Cohen, 67–86. Dordrecht: D. Reidel.

The Philosophical Novelty of Computer Simulation Methods

3.1. Introduction

Roman Frigg and Julian Reiss make two major claims in their paper "A Critical Look at the Philosophy of Simulation" (Frigg and Reiss 2009). The first is that computer simulation methods "raise few if any new philosophical problems." The second is that progress in understanding how simulations work would be enhanced by using results from the philosophical literature on models (*passim*). They are correct in their second claim. Their first claim is false and it reflects a radical failure to appreciate what is different about computational science.[1]

Here I shall lay out as clearly as I can some of the ways in which computational science introduces new issues into the philosophy of science. I shall also describe ways in which the traditional literature on models does not address these new issues and why, in many cases, it cannot. Because of space restrictions, I cannot address all of the issues that are relevant, but I hope that the essence of the philosophical points will be clear. I shall also focus on the criticisms that Frigg and Reiss make of my views, since the other authors mentioned in their article are better positioned to defend themselves and, rather than restrict the discussion to computer simulations, I shall discuss the more inclusive area of computational science, the subject which has been the focus of my own research.

3.2. The Main Issue

Let me put the principal philosophical novelty of these methods in the starkest possible way: computational science introduces new issues into the philosophy

---

[1] Stöckler 2000 argues that computer simulations do not represent a revolution in methodology. Schweber and Wächter 2000 contend that computational methods constitute a "Hacking revolution." I provide a different interpretation in Humphreys 2008.

of science because it uses methods that push humans away from the center of the epistemological enterprise. Until recently, the philosophy of science has always treated science as an activity that humans carry out and analyze. It is also humans that possess and use the knowledge produced by science. In this, the philosophy of science has followed traditional epistemology which, with a few exceptions such as the investigation of divine omniscience, has been the study of human knowledge. Locke's *Essay Concerning Human Understanding*, Berkeley's *A Treatise Concerning the Principles of Human Knowledge*, Hume's *A Treatise of Human Knowledge*, Reid's *Essays on the Intellectual Powers of Man* are but a few examples; the Cartesian and Kantian traditions in their different ways are also anthropocentric.[2] In the twentieth century, the logical component of logical empiricism broke free from the psychologism of earlier centuries, but the empiricist component prevented a complete separation.[3] Two of the great alternatives to logical empiricism, Quine's and Kuhn's epistemologies, are rooted in communities of human scientists and language users. Even constructive empiricism and its successor, the empirical stance, are firmly anchored in human sensory abilities (van Fraassen 1980, 2004). There are exceptions to this anthropocentric view, such as Popper 1972 and Ford, Glymour, and Hayes 2006, but the former's World 3 is too abstract for our concerns and the latter's artificial intelligence orientation does not address the central issues of computational science.

At this point I need to draw a distinction. Call the situation within which humans deal with science that is carried out at least in part by machines the *hybrid scenario* and the more extreme situation of a completely automated science the *automated scenario*. This distinction is important because in the hybrid scenario, one cannot completely abstract from human cognitive abilities when dealing with representational and computational issues. In the automated scenario one can, and it is for me the more interesting philosophical situation, but in the near term we shall be in the hybrid scenario and so I shall restrict myself here to that case.

For an increasing number of fields in science, an exclusively anthropocentric epistemology is no longer appropriate because there now exist superior, nonhuman, epistemic authorities. So we are now faced with a problem, which we can call the *anthropocentric predicament*, of how we, as humans, can understand and evaluate computationally based scientific methods that transcend our own abilities. This predicament is different from the older philosophical

---

[2] A Kantian approach can be generalized to nonhuman conceptual categories, although the extent to which humans could understand those alien categories is then a version of one philosophical challenge faced by computational science.

[3] Carnap's *Aufbau* ([1928] 1967) allows that a physical basis could be used as the starting point of the reconstruction procedure, but adopts personal experiences as the autopsychological basis. The overwhelming majority of the literature in the logical empiricist tradition took the human senses as the ultimate authority.

problem of understanding the world from a human perspective because the older problem involves representational intermediaries that are tailored to human cognitive capacities. With the hybrid situation, the representational devices, which include simulations, are constructed to balance the needs of the computational tools and the human consumers. This aspect of computational science is nowhere mentioned by Frigg and Reiss yet it constitutes a major epistemological change that has significant consequences of a squarely philosophical nature for how we view theories, models, and other representational items of science.

In responding to a claim about a lack of philosophical novelty, two things have to be agreed upon. One is a criterion for what counts as philosophical; the other is a criterion for what counts as new. The only way that I can address the first issue is to draw parallels between existing issues that are generally agreed to be philosophical and those raised by computational science. Of course, this is an imperfect approach just because computational science introduces philosophical problems that may be distinctively different in kind from earlier problems, but I have no other answer to the question "What counts as philosophical?" Concerning novelty, the reader will have to be the judge of what follows, but one has to keep in mind that nothing in philosophy is ever entirely new, in the sense that we can always trace a path of intellectual connections from a given position to earlier positions and that any serious claim in philosophy will have connections with other philosophical issues. The philosophical consequences of computer simulation methods are of course related to other developments in science and mathematics. There are issues about how computer-assisted mathematics changes what is acceptable as a mathematical result, although in terms of priority the use of computer simulations antedates the use of computer-assisted proofs. There is a small body of philosophical literature, mostly written by mathematicians, assessing the epistemological impact of computer-assisted mathematics and automated theorem proving. (See, e.g., Bailey and Borwein 2005; Thurston 1994.)[4] Scientific instruments that are quite different in their operation from optical microscopes and telescopes, such as radio telescopes and scanning tunneling microscopes, have been in use for decades. They too introduce philosophical challenges because these instruments have outputs that must be processed by other instruments, rather than having an output that is directly accessible to the human senses. Each of these philosophical issues is connected to philosophical aspects of computational science but the latter do not reduce to the former.

---

[4] It was unfortunate that the early philosophical discussions of computer-assisted mathematics became distracted by focusing on the issue of whether a different sense of proof was involved. Perhaps this is why a recent 800-page reference work on the philosophy of mathematics (Shapiro 2005) has no discussion of computer-assisted mathematics.

3.3. What Is Philosophically New about Computational Science

There are at least four specific philosophical issues related to the anthropocentric predicament which the study of computational science requires us to address. These are the essential epistemic opacity of most processes in computational science, the relations between computational representations and applications, the temporal dynamics of simulations, and the need to switch from in-principle results to in-practice considerations. Those four issues form a connected set and each will be treated in some detail below. I start with epistemic opacity.

3.3.1. EPISTEMIC OPACITY

One of the essential features of computational science, which is not mentioned by Frigg and Reiss, is the essential epistemic opacity of the computational process leading from the abstract model underlying the simulation to the output. Here a process is epistemically opaque relative to a cognitive agent X at time t just in case X does not know at t all of the epistemically relevant elements of the process. A process is essentially epistemically opaque to X if and only if it is impossible, given the nature of X, for X to know all of the epistemically relevant elements of the process.[5] For a mathematical proof, one agent may consider a particular step in the proof to be an epistemically relevant part of the justification of the theorem, whereas to another, the step is sufficiently trivial to be eliminable. In the case of scientific instruments, it is a long-standing issue in the philosophy of science whether the user needs to know details of the processes between input and output in order to know that what the instruments display accurately represents a real entity.

Within the hybrid scenario, no human can examine and justify every element of the computational processes that produce the output of a computer simulation or other artifacts of computational science. This feature is novel because, prior to the 1940s, theoretical science had not been able to automate the process from theory to applications in a way that made the details of parts of that process completely inaccessible to humans.[6] Many, perhaps all, of the

---

[5] In my 2004, I used only the straightforward "epistemically opaque" terminology. I now think that distinguishing between the weaker and stronger senses is useful. It is obviously possible to construct definitions of "partially epistemically opaque" and "fully epistemically opaque," which the reader can do himself or herself if so inclined. What constitutes an epistemically relevant element will depend upon the kind of process involved.

[6] I would put the two important turning points in the mid-1940s, when Monte Carlo methods were first implemented on electronic computers and John Mauchly suggested that ENIAC could be used for difference equation calculations, rather than for just routine arithmetical operations. On the latter claim, see Metropolis 1993, 127. I do not vouch for the accuracy of Metropolis's recollections on this point, although the exact historical turning point, if indeed "exact" ever makes sense in historical claims, is unimportant.

features that are special to simulations are a result of this inability of human cognitive abilities to know and understand the details of the computational process. The computations involved in most simulations are so fast and so complex that no human or group of humans can in practice reproduce or understand the processes. Although there are parallels with the switch from an individualist epistemology, within which a single scientist or mathematician can verify a procedure or a proof, to social epistemology, within which the work has to be divided between groups of scientists or mathematicians, so that no one person understands all of the process, the sources of epistemic opacity in computational science are very different.

I can illustrate some of the issues involved using as an example agent-based simulations, which were only briefly discussed in my 2004. The brevity of that discussion is a major deficiency of the book and the relative lack of attention I gave those methods can be misleading because agent-based simulations are in certain ways very different from what one might call equation-based simulations. It is a common, although not universal, feature of agent-based models that emergent macro-level features appear as a result of running the simulation, that these features would not appear without running the simulation, that new macro-level descriptions must be introduced to capture these features, and that the details of the process between the model and its output are inaccessible to human scientists. No traditional modeling methods address the first, second, and fourth features of these simulations. Let me elaborate a little on this point. The situation has been nicely captured by Stephen Weinberg: "After all, even if you knew everything about water molecules and you had a computer good enough to follow how every molecule in a glass of water moved in space, all you would have would be a mountain of computer tape. How in that mountain of computer tape would you ever recognize the properties that interest you about the water, properties like vorticity, turbulence, entropy, and temperature?" (Weinberg 1987, 434). Many of these "higher level" conceptual representations already exist in other theoretical representations; they are the starting point for what Ernest Nagel called inhomogeneous reductions. With other agent-based models the situation is different because the simulation itself will, in some cases, construct a novel macro-level feature. These emergent patterns in computer simulations form the basis for what Mark Bedau has characterized as "weak emergence" (Bedau 1997) and traditional human modeling techniques will not generate them from the agent base. They can only be arrived at by simulation.

3.3.2. SEMANTICS

Philosophy of science has, as one of its concerns, how theories, models, and other representational devices are applied to real systems. One of the most common sources of frustration that scientists express about the philosophy of science is its detachment from the often brutal realities of getting theory into contact with data; how a scientific representation is applied to a real system involves considerably more than is included in the traditional semantical concerns of reference, meaning, and truth. Although many philosophers will insist that providing a semantics for a simulation is necessary for it to be applied, and in the hybrid situation that is true, semantics and application are different because one can have a theory with a fully specified semantics that cannot be applied.[7]

Frigg and Reiss agree that the issue of application involves more than just semantics, but they almost immediately dismiss the most important aspect of how simulations are related to their applications. They say, "In the broadest sense of application—meaning simply the entire process of using the model . . . we use computational methods rather than paper and pencil to get the solutions of the equations that form part of the model. But if this is the claim, then this is just a restatement in 'application jargon' of the point of departure, namely that there are equations which defy analytical methods and have to be solved numerically." The slide from "computational methods" to "equations . . . that have to be solved numerically" seriously underdescribes how computational science deals with the application task. Each of the other three issues mentioned earlier—opacity, dynamics, possibility in practice—is relevant to the process of computationally applying a scientific representation to a real system.

In Humphreys 2004, sec. 3.12 I argued that syntax was an important element of computer simulations but that neither the syntactic account of theories nor the semantic account of theories was a suitable vehicle for representing simulations. Frigg and Reiss say that my claim is puzzling because "simulations by themselves do not clash with either the semantic or the syntactic view." The point is that both of these accounts of theories lack the resources to capture the essential features of computational science. The semantic account of theories, which requires that one abstract from the syntactic representation of the theory, is the wrong vehicle for capturing simulations because the specific syntactic representation used is often crucial to the solvability of the theory's equations. Syntax is also inseparable from the dynamic implementation of simulations. Computer simulations, because they are essentially dynamic processes taking place in time, involve the processing of linguistic strings on concrete machines. The only way to run a simulation is to run the code. Of course there are abstract representations of the algorithms involved, but it is a category mistake to claim that a collection of code in a file is a simulation. The

---

[7] Paul Teller has argued that simulations cannot be treated as completely formal objects because of the problem of intentionality. (APA Pacific Division meetings, spring 2005, unpublished talk.)

program has the disposition to produce a simulation when run on a suitable machine but it only becomes a simulation when it is running.

However, we require more than the syntactic account of theories provides because the way in which simulations are applied to systems is different from the way in which traditional models are applied. As we described above, there is the issue of how, abstractly, the syntax is semantically mapped onto the world, on which the traditional syntactic account has a great deal to say, and then there is the different issue of how the theory is actually brought into contact with data. It is in replacing the explicitly deductive relation between the axioms and the prediction by a discrete computational process that is carried out in a real computational device that the difference lies.

Frigg and Reiss do occasionally address the specifically computational aspects of computational science, but when they do, they claim that the problems are of a mathematical nature, not philosophical. The basis of their claim cannot be that mathematical results and techniques do not have philosophical consequences, because this is obviously false. The invention of non-Euclidean geometries, Gödel's incompleteness results, the initial proof of the four-color theorem, renormalization theory in mathematical physics, and the differences between classical and Bayesian statistical inference methods are only a few examples of how developments in mathematics can give rise to significant philosophical issues. So the claim must be, once again, that no new philosophical issues are raised by technical issues in computational science.

Frigg and Reiss present three arguments against the need for a new relation between the representation and the system. The first appeals to the example of two pendulums, one a normal pendulum having a computationally tractable model and the other a double pendulum with a model that can only be solved numerically. They then ask, "Does this change [from tractable to intractable mathematics] change our understanding of how the equation relates to the world?" and answer in the negative: "Nothing in our empirical interpretation of the terms of the equation changes in any way." This response rests on accepting that the task is the narrow one of providing a semantics for the syntax, rather than the broader task of how that syntax is applied, and I have argued above that simulations connect models to applications in ways that go well beyond semantics.

The second argument presents a more serious challenge. One of the major unresolved issues in many areas of computational science is whether the invention of new mathematical techniques might eventually replace some of these computational methods. Frigg and Reiss suggest that if we introduced a new class of functions that were solutions to the existing, currently intractable model, this would not change the way the model relates to the world. In fact it would, because with the availability of analytic solutions, the epistemic opacity of the relation between the model and the application would disappear. Moreover, even if this were to happen, the fact that the computational methods are, during our era, an unavoidable part of scientific method makes them of philosophical interest, just as the use of the Ptolemaic apparatus for computing planetary orbits is still of philosophical interest.

Their third argument is that, given the first two arguments, "whether simulations make either the syntactic or the semantic view of theories obsolete becomes a non- issue." This does not follow because there are aspects of computational science that are simply not addressed by either of the two philosophical accounts of theories. The traditional syntactic account of theories distinguished between some types of theories; those that were recursively axiomatizable, those whose axioms sets are only recursively enumerable, and a few other types. Computer scientists have since added to this classification, in moving from the simple issue of (Turing) computability to measures of theoretical computational complexity, such as P, NP, P-SPACE, and many others. This refinement can be incorporated within the syntactic account of theories. Other issues about the power of different computational architectures that are also relevant to computational science cannot be so incorporated. It is possible that if operational quantum or biological computers are built, a number of scientifically intractable problems will become tractable, opening up new areas of research. This is not an issue that is in any way addressed by traditional modeling techniques and although philosophical discussions of quantum computing have not been motivated much by issues in the area of simulations, the area is novel and is relevant to computational science. (See Mermin 2007.)

3.3.3. TEMPORAL DYNAMICS

One of the features of computer simulations, although by no means the only one, that produces essential epistemic opacity is the dynamic, temporal nature of the computational process. Because in section 1 of their article, Frigg and Reiss set aside Stephan Hartmann's dynamic characterization of simulations, to which I subscribe in a rather different form, it is not surprising that they fail to address this issue.[8] The issues that Frigg and Reiss bring out about the dynamics of simulations are important and this is an appropriate place to elaborate on my earlier treatment. In Humphreys 2004 I distinguished between a core simulation and a full representation:

> System S provides a core simulation of an object or process B just in case S is a concrete computational device that provides, via a temporal process, solutions to a computational model in the sense of Sect. 3.14 [of Humphreys 2004] that correctly represents B, either dynamically or statically.

---

[8] Their setting aside Hartmann's account on the grounds that "those who put forward the claims at issue here do not use this definition" is startling, since section 4.1 of my 2004 is devoted to discussing and endorsing a version of Hartmann's account that retains an essential dynamic element.

If in addition the computational model used by S correctly represents the structure of the real system R, then S provides a core simulation of system R with respect to B. . . . In order to get from a core simulation to a full simulation, it is important that the successive solutions which have been computed in the core simulation are arranged in an [appropriate] order . . . the process that constitutes the simulation consists of two linked processes—the core process of calculating the solutions to the model, within which the order of calculations is important for simulations of the system but unimportant for simulations of its behavior, and the process of presenting them in a way that is either a static or a dynamic representation of the [behavior of the real system], within which the order of representation is crucial. (2004, 110)

There are two distinct roles played by time in full simulations. The first is the temporal process involved in actually computing the consequences of the underlying model. The second is a temporal representation of the dynamical development of the system. The first process is always present in computational science. Depending on the application, the second may be present or absent, as it will be in a static representation of the system's states. Frigg and Reiss recognize this double role played by time. Traditional, abstract representations of models and their relation to applications do not contain the temporal element involved in core simulations because they abstract from the implementation level. Here we can see another way in which philosophical assessments of computational science are different from the traditional modeling literature. When logicians deal with measures of computational complexity, it is how fast the number of computational steps grows relative to the input size that is important. These measures abstract from the actual time taken on a real machine and this is an appropriate abstraction for a logician. But for philosophers of science, the very nature of a prediction, as opposed to a deduction, rests on the ability to produce the result temporally in advance of the state of affairs being predicted. A model of weather dynamics would be useless for predictive purposes if any simulation based on it would take 10^6 years to run. It is because of this that in computational science technological considerations cannot be separated from philosophical considerations.

We can now see that the alternative suggestion made by Frigg and Reiss that "the actual computational steps and the time they take to be executed somehow represent the time that the processes in the world take to unfold" (10) fails to respect the distinction between the two roles played by time in simulations. Even with a static representation of the system's states, the core simulation will still involve a temporal dimension. There is no reason to require that the core simulation bears a mimetic relation to the system's temporal development— that is why a separate consideration of the output representation is crucial for a full simulation. The decision about the output representation is important in simulations; for example the difference between synchronous and asynchronous updating schedules, which decide how often parts of the core simulation are used to update the system state, can make a crucial difference in the results obtained from agent-based models. The principal philosophical point, however, is that the epistemic opacity, the dynamic aspects, the nature of the application process, and the need to pay attention to what is possible in practice all depend on the real temporal nature of the core simulation.

3.3.4. IN PRACTICE, NOT IN PRINCIPLE

A fourth novel feature of computational science is that it forces us to make a distinction between what is applicable in practice and what is applicable only in principle. Here the shift is, first, from the complete abstraction from practical constraints that is characteristic of much of traditional philosophy of science and, second, from the kind of bounded scientific rationality that is characteristic of the work of Simon and Wimsatt, within which the emphasis tends to be on accommodating the limitations of human agents. Ignoring implementation constraints can lead to inadvisable remarks. It is a philosophical fantasy to suggest, as Frigg and Reiss do, that "If at some time in the future we have a computer that can calculate all the states in no time at all" or, as Stöckler does, that "In principle, there is nothing in a simulation that could not be worked out without computers" (2000, 368).[9]

In saying this I am not in any way suggesting that in principle results are not relevant in some areas. They clearly are; there are also other issues to which the philosophy of science needs to devote attention. One of the primary reasons for the rapid spread of simulations through the theoretically oriented sciences is that simulations allow theories and models to be applied in practice to a far greater variety of situations. Without access to simulation, applications are sometimes not possible; in other cases the theory can be applied only to a few stylized cases.

Within philosophy, there is a certain amount of resistance to including practical considerations, a resistance with which I can sympathize and I am by no means suggesting that the investigation of what can (or cannot) be done in principle is always inappropriate for the philosophy of science. One source of resistance to using in practice constraints is already present in the tension between descriptive history of science and normative philosophy of science, and in the tension between naturalistic approaches (which tend to mean different things to different people) and more traditional philosophy of science. But the appeal to in-principle arguments involves a certain kind of idealization, and

---

[9] The first versions of Thomas Schelling's agent-based models of segregation, and the first versions of Conway's Game of Life were done "by hand," but almost all contemporary simulations require abilities that go far beyond what is possible by the unaided human intellect.

some idealizations are appropriate whereas others are not. A long-standing
epistemological issue involves the limits of knowledge. Are there things that
we cannot know, and if so, can we identify them? There surely cannot be any
issue that this is a genuine philosophical problem. Of course it is not new—
Kant famously gave us answers to the question. The question of what we can
know in philosophy of science has been transformed by the rise of computational science and it is partly a question of what idealizations can legitimately
be used for epistemic agents. We already have experience in what idealizations
are appropriate and inappropriate for various research programs. The move
away from hyperrational economic agents in microeconomics to less idealized
agents within behavioral and experimental economics is one well-known example. For certain philosophical purposes, such as demonstrating that some
kinds of knowledge are impossible even in principle, in-principle arguments
are fine. But just as humans cannot in principle see atoms, neither can humans
in principle compute at petaflop speeds.

3.3.5. OTHER ISSUES

Frigg and Reiss claim, correctly, that much of what goes into constructing
the models that lie behind many computer simulations involves issues that
have been discussed in the previous philosophical literature on models. They
write: "But again, approximations, simplifications, idealisations, and isolations
are part and parcel of all science and in no way specific to the use of computers
in science." True enough, but where is the list of things that *are* specific to
computer simulations? These include constraints put on models by computational load issues, the problems of extending models when substantial chunks
of existing code are written in legacy software, the choice of finite element decomposition, and the need for research teams to delegate substantial amounts
of authority to programmers, to name just a few. Philosophers of science are
free to abstract from all of these issues, but then in some areas of science their
accounts will simply misrepresent how progress is made.

Even with idealizations, these computational features are relevant. Here is
one particular example: Determining energy levels is a core interest for molecular chemists. Physical chemistry employs quantum mechanics as its basic
theoretical apparatus, but ab initio calculations of the energy levels are impossible to carry out for any but the smallest molecules. The simple valence
bond and molecular orbital models do not provide accurate predictions even
for hydrogen molecules, so they have to be supplemented with dozens of
extra terms to account for various features. They therefore employ multiple
approximations and are heavily computational. So the approximations chosen
in the Hartree-Fock self-consistent field approach, a central method of computational quantum chemistry, are inextricably linked with the degree to which
those calculations can actually be carried out in practice. On the other side,

in 2004, I focused on the constraints that restricted computational resources place on computational science. There is now a growing sense that a different problem has arisen; that new techniques need to be developed to effectively exploit the massive computational power that is now available in many areas.[10]

3.4. Conclusion

Frigg and Reiss attribute four claims to the contemporary philosophical literature on simulations, each of which they argue is either wrong or not new:

 (a) A metaphysical claim: "Simulations create some kind of parallel world in which experiments can be conducted under more favorable conditions than in the 'real world.'"
 (b) An epistemological claim: "Simulations demand a new epistemology."
 (c) A semantic claim: "Simulations demand a new analysis of how models/ theories relate to concrete phenomena."
 (d) A methodological claim: "Simulating is a *sui generis* activity that lies 'in between' theorizing and experimentation."

I would add at least one more: A fifth aspect of simulations is that in the mathematically oriented sciences, progress is now inescapably linked to technological progress. Frigg and Reiss's claim about the metaphysical consequences of simulations is essentially correct and I have never subscribed to that metaphysical position myself. I have argued in this article that their second and third claims are incorrect. Computational science requires a new nonanthropocentric epistemology and a new account of how theories and models are applied. These requirements are, to me, more than sufficient to justify the claim that computational science is a significantly new sui generis activity accompanied by new, recognizably philosophical, issues. Claims that these methods lie "in between" theorizing and experimentation are, I believe, best interpreted metaphorically. The phrase indicates that computer simulations often use elements of theories in constructing the underlying computational models and they can be used in ways that are analogous to experiments. (For details of one of these ways, see Humphreys 1994.) Computational science has also made possible almost everything that takes place in complexity theory, itself a new area of science with its own methods that has powerful cross-disciplinary capabilities; classes dedicated to computational physics, chemistry, and biology along with textbooks dedicated to the topics have been introduced in those departments

---

[10] "Rationale for a Computational Science Center," unpublished report, University of Virginia, March 2007.

because the methods involved are different from those taught in their theory classes and in laboratory sessions.

Frigg and Reiss make many valuable points in their article; indeed, among the skeptical literature in this area, their arguments are the clearest that I have encountered. Nevertheless, these powerful new currents sweeping through the sciences bring with them philosophical challenges that older modeling frameworks cannot address. It is not a matter of lightly abandoning successful methods but of adapting to a different world.

References

Bailey, David, and Jonathan M. Borwein. 2005. "Future Prospects for computer assisted mathematics." *Notes of the Canadian Mathematical Society* 37: 2–6.

Bedau, Mark. 1997. "Weak Emergence." *Philosophical Perspectives* 11: 375–99.

Carnap, Rudolf. (1928) 1967. *Der logische Aufbau der Welt*. Berlin. Translated by Rolf George as *The Logical Structure of the World*. Berkeley: University of California Press.

Ford, Kenneth, Clark Glymour, and Patrick Hayes. 2006. *Thinking about Android Epistemology*. Menlo Park, CA: AAAI Press.

Frigg, Roman, and Julian Reiss. 2009. "The Philosophy of Simulation: Hot New Issues or Same Old Stew?" *Synthese* 169: 593–613.

Humphreys, Paul. 1994. "Numerical Experimentation." In *Patrick Suppes: Scientific Philosopher*, vol. 2, *Philosophy of Physics, Theory Structure and Measurement Theory*, edited by Paul Humphreys, 103–18. Dordrecht: Kluwer Academic.

———. 2004. *Extending Ourselves: Computational Science, Empiricism, and Scientific Method*. New York: Oxford University Press.

———. 2008. "Computational Science and Its Effects." ZiF Mitteilungen, Zentrum für interdisziplinäre Forschung, Bielefeld. Reprinted as Chapter 9 of *Science in the Context of Application*, Boston Studies in the Philosophy of Science 274, edited by Martin Carrier and Alfred Nordmann, 131–142. Berlin: Springer, 2011.

Mermin, N. David. 2007. *Quantum Computer Science*. Cambridge: Cambridge University Press.

Metropolis, Nicholas. 1993. "The Age of Computing: A Personal Memoir." In *A New Era in Computation*, edited by Nicholas Metropolis and Gian-Carlo Rota, 119–30. Cambridge, MA: MIT Press.

Popper, Karl. 1972. "Epistemology without a Knowing Subject." In *Objective Knowledge: An Evolutionary Approach*, edited by Karl Popper, 106–52. Oxford: Oxford University Press.

Schweber, Sam, and Matthias Wächter. 2000. "Complex Systems, Modeling and Simulation." *Studies in History and Philosophy of Modern Physics* 31: 583–609.

Shapiro, Stewart, ed. 2005. *The Oxford Handbook of Philosophy of Mathematics and Logic*. New York: Oxford University Press.

Stöckler, Manfred. 2000. "On Modeling and Simulations as Instruments for the Study of Complex Systems." In *Science at Century's End: Philosophical Questions on the Progress and Limits of Science*, edited by Martin Carrier, Gerald Massey, and Laura Ruetsche, 355–73. Pittsburgh: University of Pittsburgh Press.

Thurston, W. 1994. "On Proof and Progress in Mathematics." *Bulletin of the American Mathematical Society* 30: 161–77.

van Fraassen, Bas. 1980. *The Scientific Image*. Oxford: Clarendon Press.

———. 2004. *The Empirical Stance*. New Haven: Yale University Press.

Weinberg, Stephen. 1987. "Newtonianism, Reductionism, and the Art of Congressional Testimony." *Nature* 330: 433–37.

{ 4 }

Numerical Experimentation

4.1

Over the course of the past couple of decades, computational science has become an increasingly important method in many areas of the physical, biological, psychological, economic, and other sciences. Because of the wide variety of methods subsumed under the general framework of computational science, it is not easy to characterize the methods succinctly, but for the purposes of this paper I shall use the following working definition:

> Computational science is the development, exploration and application
> of mathematical models of nonmathematical systems using concrete
> computational devices.

To flesh out this working definition, I include in the class of concrete computational devices both digital and analog computers (i.e., analog devices that are used for computational purposes and not just as physical models). Humans are one special kind of computational device, but the kind of novel philosophical issues that I want to discuss here occur only when we move beyond the category of human computers. Moreover, it is important that the computation is actually carried out on a concrete device, for mere abstract representations of computations do not count as falling within the realm of computational science. To be counted as part of computational science proper, the computations must have a dynamic physical implementation. (I do, however, include computations run on virtual machines in the class of concrete computations.)

This working definition is simple, but it already subsumes such procedures as applied finite difference methods, Monte Carlo methods, molecular dynamics, Brownian dynamics, semiempirical methods of computational chemistry, and a whole host of other, less familiar methods. For those of us who are interested in how theories are applied to nature, the most important immediate effect of using these methods of computational science is that a vastly increased number of models can be brought into significant contact with real systems, primarily by circumventing the serious limitations that our present restricted set of analytic mathematical techniques imposes on us. These constraints should not be underestimated, for once we move past the realm of highly idealized systems, the number of mathematical theories that are applicable using only analytic techniques devoid of approximations is very small. I shall not dwell on this aspect here, but refer the reader to Humphreys 1991 for examples of this analytic unsolvability.

Many of the other features of computational science have no real interest for philosophy, important though they are to the working scientist. But some aspects of this methodology give rise to questions that have clear philosophical content. Among these I shall focus on two specific issues. First, it is often claimed that computational science provides us with a new kind of scientific method, one that is complementary to the existing methods of theory and (empirical) experimentation. Such claims are often rather vague about just what is new in these methods, ranging through a variety of claims such as (1) that computational science allows both a reduction of the degree of idealization needed in the models as well as a check on those remaining idealizations that are made; (2) that simulations allow far more flexibility and scope for changing boundary and initial conditions than do real experiments; (3) that simulations are precisely replicable; (4) that many parameters can be varied that could not be altered by real experiments, perhaps because such a change would violate a law of nature, perhaps because as in the case of astrophysical systems the very idea of large-scale experiments is absurd, and so on.

I shall argue here that at least one specific aspect of computational science does indeed introduce a genuinely novel kind of method into science. That method is numerical experimentation, and I shall argue for this claim with reference to a particular class of mathematical models—the lattice models of statistical physics, using a particular type of computational method—the Monte Carlo method—and applying a specific solution procedure to that method—the Metropolis algorithm. Because of the wide applicability of Monte Carlo methods within computational science, this example constitutes a lesser degree of special pleading than might appear at first sight. Fritz Rohrlich 1991, following Naylor 1966 and others, has similarly argued for the importance of computer simulations as a kind of "theoretical model experimentation." The examples used here reinforce his perspective while exhibiting a somewhat different aspect of computational experimentation, and they have the additional advantage that they tie together two topics of perennial philosophical concern, probability and empiricism, in a rather unusual way.

The second feature of philosophical interest involves a parallel that is sometimes drawn between the use of computational science and the use of scientific instruments. Whereas the latter enable us to transcend the limitations of our unaided sensory capacities, the former, so it is claimed, enable us to extend our limited mathematical abilities, especially in cases where computational processes play an important role. The degree to which this parallel holds, if at all, is important at least because the instrument side of the parallel has had profound and well-known effects on the development of contemporary scientific empiricism. We are all familiar with the arguments that were developed in the middle third of the twentieth century which produced serious difficulties for certain kinds of foundationally inspired empiricisms. Many things too small or too far way to be seen with the naked eye and things emitting radiation outside the realm of the visible are now routinely considered to be observable. For example, observing a cold virus under an electron microscope seems to many of us to be not only a perfectly legitimate part of scientific practice, but one that ought to be acceptable to a liberal empiricist. There is no need to rehearse here the consequences that extending the concept of "observable" to include what is detectable by scientific instruments has had on empiricism. It is enough to remind oneself that they have been profound. So we must ask: if the parallel between computational science and instrumentation is soundly based, what consequences are there for how we should construe the use of mathematical models in science? I shall restrict myself here to the issue of how the parallel affects numerical experimentation, and in particular, what kinds of epistemological criteria are appropriate to the new methodology. But enough of the preliminaries. Let us turn to the details of the model we shall examine.[1]

4.2

Ferromagnetism involves spontaneous magnetization of materials such as iron and nickel when the temperature is lowered below the Curie or critical temperature T_c. The exchange energy involved in ferromagnetism is a specifically quantum mechanical phenomenon manifesting itself at the macroscopic level, and one of the central applications of the lattice models is to study this phenomenon.[2] These models consist of an m-dimensional lattice ($m \leq 3$) with the nodes occupied by particles having spin values S_i. The Hamiltonian for the Ising model, which describes a ferromagnet with strong uniaxial anisotropy is given by

$$\mathcal{H}_{ISING} = -J \sum_{ij} S_i S_j - \mu H \sum_i S_i$$

---

[1] Those uninterested in the technical details of these models can move directly to section 4.4, referring back as necessary. It is not possible to fully understand why numerical experimentation is forced on us without a grasp of the physical models, however.

[2] The exposition of the formal aspects of lattice models given here relies heavily on the treatment in Binder and Heerman 1988.

Here J is the exchange energy constant, restricted initially to interaction between nearest neighbors. If $J > 0$, then the pairs (up,up), (down,down) are favored for neighbor pairs and this gives rise to the spin alignments that produce ferromagnetism. If $J < 0$, then the pairs (up,down), (down,up) are favored, resulting in antiferromagnetism. H is the external magnetic field, μ is the magnetic moment of a spin, and the second term on the right is the Zeeman energy. When generalized to a fully isotropic ferromagnet in the Heisenberg model we obtain

$$\mathcal{H}_{HEISENBERG} = -J \sum_{ij} (\mathbf{S}_i \cdot \mathbf{S}_j) - \mu H_z \sum_i S_i^z,$$

where

$$(S_i^x)^2 + (S_i^y)^2 + (S_i^z)^2 = 1$$

Here, as opposed to the Ising model in which the spins have values $+1$ or -1 along the preferred axis, in the Heisenberg model the spins can take on continuously many orientation values. Among the simplifying assumptions used in these models are (a) the kinetic energy of the particles associated with the lattice site is neglected; (b) only nearest neighbor interactions are considered (although this can be relaxed to include nth neighbor interactions); (c) spins have only two discrete values; (d) J, H are considered to be uniform (this can also be modified to allow random exchange constants J_{ij} and different magnetic fields at each point in the lattice).

Contact with empirical data comes through the calculation of average values for macroscopically accessible quantities, such as the mean magnetization. In order to calculate these averages, the model is supplemented by a choice of thermodynamical ensemble with its associated probability distribution over states of the system. A common choice here is the canonical ensemble, i.e., an ensemble of closed systems in thermal equilibrium with a heat reservoir. Then for a (macroscopic) observable A and a system with continuously many degrees of freedom, the thermal average is given by

$$< A(\mathbf{x}) >_T = Z^{-1} \int_X exp\left[-H(\mathbf{x})/k_B T\right] A(\mathbf{x}) \, d\mathbf{x}, \qquad (1)$$

where

$$Z = \int_X exp\left[-H(\mathbf{x})/k_B T\right] d\mathbf{x}$$

In the isotropic Heisenberg model, we have continuous degrees of freedom leading to a proper integral, but in models where there are only finitely many degrees of freedom, such as the Ising model, the integral in (1) will, of course, be replaced by a summation. The integration or summation is over the configuration space X, which in the lattice models has as points N-dimensional vectors $\mathbf{X} = (S_1, \ldots, S_N)$ specifying the spin state for each particle in the lattice. This introduces in an explicit way one probabilistic aspect of the model in that the normalized Boltzmann distribution $exp[-H(\mathbf{x})/k_B T]/Z$ describes the probability of the configuration $\mathbf{x}$ in thermal equilibrium.

The representation given in equation (1) is familiar, tidy, and explicit. Yet even given the severe simplifications involved in (a) through (d) above, equation (1) is impossible to apply in any realistic context. The integrals involved in the Heisenberg model are unsolvable in any explicit way, and although there exist analytic solutions to the Ising model for one- and two-dimensional lattices, there is no such analytic treatment for the three-dimensional lattice. Moreover, even in the discrete case, for very simple models of lattices with 10^2 nodes in each spatial dimension, there are $2^{1000000}$ possible states to sum over in the three-dimensional case. Analytically intractable integrals produce one standard context in which Monte Carlo methods are used, and in employing these methods we switch from a treatment in which a determinate solution is produced by analytic methods to one in which a statistical estimate of that solution is provided. To be clear about how probabilities operate in this context, it is essential to separate the fact that we began with a probabilistic model involving a probability distribution for the canonical ensemble over configuration space [$\mathbf{x}$] from the probabilistic techniques that are used to produce a solution to the multiple dimensional integral (1). We are here concerned with the second of these probabilistic aspects. Traditional integration methods, including those using approximation methods, do not treat these averages as stochastic quantities to be estimated by probabilistic methods. In contrast, Monte Carlo methods do just that, and in fact Monte Carlo methods can be applied equally well to solve models that are based on processes that are themselves deterministic. A brief outline of how these methods are applied is needed here.

4.3

The bridge between traditional methods of integration and Monte Carlo methods is provided by the mean value theorem, given here for the one-dimensional case: if u, v are continuous functions on $[a, b]$ and v does not change sign in the interval $[a, b]$, then there is an ξ in $[a, b]$ such that

$$\int_a^b u(x)v(x)\,dx = u(\xi)\int_a^b v(x)\,dx$$

Thus, when $v(x) = 1$ everywhere, we have

$$\int_a^b u(x)\,dx = u(\xi)(b-a). \tag{2}$$

Here $u(\xi)$ is the mean value of u on $[a, b]$.

Hence the point of Monte Carlo methods is to statistically estimate the mean value $u(\xi)$, then to calculate

$$\int_a^b u(x)\, dx$$

using (2). Now suppose we consider the quantity $u(X)$ to be a random variable, and we sample values $u_i = u(X_i)$ at random. Then

$$\overline{u(x)} = \sum_{i=1}^{N} u(X_i)\,/\,N \tag{3}$$

Finally, by appeal to the strong law of large numbers, one can justify probabilistic convergence of the statistical estimator $u(x)$ to the analytic mean value $u(\xi)$.

Now this argument clearly has nothing to do with computational devices; it is a result that derives solely from familiar facts about the calculus and probability theory. But there are two facts that make this otherwise routine piece of numerical mathematics potentially interesting. The first is that because of the large dimension of the configuration space involved in the lattice models (and in most other models with any pretense to realism) the computation involved in (3) has to be carried out in practice by resorting to computational devices using the technique that is our principal focus of attention here, *numerical experimentation*. This technique can be illustrated by means of the Metropolis algorithm. Within the general procedures for Monte Carlo estimators, drawing sample points from configuration space for multidimensional integrals in a random way often results in large portions of the configuration space being sampled negligibly often. Using a uniform probability distribution on the points of configuration space is usually inefficient, because many points of the space are highly improbable and correspond to negligibly small contributions to the estimator. Given the large dimensional space over which the estimation takes place, this will produce a rate of convergence that is far too slow for the computation to be carried out in practice. If, instead, the points in configuration space are selected by a distribution that mimics the profile of the function that is already being used to weight the quantity whose mean we are interested in, this makes much more efficient use of the sampling procedure. To see this, suppose we have chosen an ensemble (here the canonical ensemble) that has a probability distribution $f(H(\mathbf{x}))$ over the energy states. Then, in *simple* sampling, we would approximate the ensemble average for an observable A by

$$<A> \approx \sum_{i=1}^{M} A(\mathbf{x}_i) f(H(\mathbf{x}_i))\,/\,\sum_{i=1}^{M} f(H(\mathbf{x}_i))$$

where $\mathbf{x}_i$ is some element from phase space (e.g., the N-dimensional spin space $[\mathbf{x}_i = (S_1, \ldots, S_N)]$). But to avoid wasting time on elements of phase space with low probability, we choose the elements with probability $P(\mathbf{x}_i)$ to get

$$< A > \approx \sum_{i=1}^{M} A(\mathbf{x}_i) P^{-1}(\mathbf{x}_i) f(H(\mathbf{x}_i)) / \sum_{i=1}^{M} P^{-1}(\mathbf{x}_i) f(H(\mathbf{x}_i)).$$

(Note: this comes from choosing points according to the measure $P(\mathbf{x}_i)$ d$\mathbf{x}$ instead of the uniform measure d$\mathbf{x}$.)

Now in the case of Ising models, suppose we choose $P(\mathbf{x}_i)$ to be proportional to the equilibrium distribution $f(H(\mathbf{x}_i))$, which here is the Boltzmann distribution. So, for example, if in the Ising model we choose a sampling distribution

$$P(x) = Z^{-1} \exp[-H(x)/k_B T]$$

then the weighted average

$$< A > \approx \frac{\sum_{i=1}^{M} A(\mathbf{x}_i) P^{-1}(\mathbf{x}_i) \exp[-H(\mathbf{x}_i)/k_B T]}{\sum_{i=1}^{M} P^{-1}(\mathbf{x}_i) \exp[-H(\mathbf{x}_i)/k_B T]}$$

reduces to

$$< A(x) > \approx \sum_{i=1}^{M} A(\mathbf{x}_i)/M.$$

But in many cases the exact values of the equilibrium distribution

$$\exp[-H(\mathbf{x}) = k_B T]$$

are not known for a specific Hamiltonian, and so we would seem to have made no progress.

To circumvent the difficulty, the Metropolis algorithm constructs a random walk in configuration space having a limit distribution identical to that of the sampling function P. If this were merely a piece of abstract probability theory about Markov processes it would, of course, be nothing exceptional. The key fact about this method is that the limit distribution is not computed by calculating a priori the values of the limit distribution. Rather, by running concrete random walks within numerical models of configuration space, the limit distribution so generated is proportional to the equilibrium distribution.

The Metropolis algorithm is easily described. First, pick an arbitrary point in configuration space, X_0. Then, if we are at point X_i, to generate a move in the random walk, choose a new point in configuration space, X_j. Compute the transition probability $w(X_i \rightarrow X_j) = r$ for which one common choice is the Metropolis function:

$$w(X_i \rightarrow X_j) = \min\{1, \exp(-[H(X_i) - H(X_j)]/k_B T)\}$$

Next, generate a uniformly distributed random number rnd, and if $r > rnd$ let $X_{n+1} = Y$, otherwise let $X_{n+1} = X_n$. Iterate the process for sufficiently many steps until equilibrium occurs. Then repeat the random walk choosing a different initial starting point. Iterate until a frequency distribution of equilibrium states has been generated by the equilibrium end states of the repeated random walk. (For a proof that $P_{eq}(X_i) \sim w(X_i)$, see Metropolis et al. 1953.)

Coupled with the transition probability is a physically interpretable dynamics in configuration space that generates new states from old. Two common choices for the Ising model are the single spin flip dynamics, within which the spin at the selected point is reversed, and the spin exchange dynamics, within which the spin at the selected point is exchanged with a neighbor. Putting this in the concrete case of spin flip dynamics for the Ising model, an initial configuration for the lattice is chosen, and then a random lattice site is picked and the spin there is flipped. If this results in a decrease in total energy, then the Metropolis transition probability (which in that case will be unity) will choose the new configuration and the next step in the random walk will be generated. If, in contrast, the spin flip results in an increase in total energy, then the Metropolis transition probability is compared to the uniformly distributed random number, and if the transition probability is greater, the new configuration is accepted. If not, the process retains the old configuration. What the Metropolis algorithm does, then, is to get from an initial configuration to one in which (a) the energy is globally, rather than locally, minimized and (b) the probabilities are distributed according to the equilibrium distribution.

4.4

There are certain distinctive features of numerical experimentation and the Metropolis algorithm that need emphasis. First, the equilibrium distribution that the algorithm has as a limit is not in general analytically computable in the sense that its values can be calculated explicitly for a closed form representation of the probability. The only way to compute these probabilities is by means of the numerical random walk generated by the Metropolis algorithm. Although this random walk is a model of a physical process, where the generation of the random numbers could in principle be carried out by physically indeterministic means, thus resulting in a traditional kind of empirical experimentation using stochastic processes, the method described here is carried out purely numerically. It is this simulation by numerical methods of physical process that has led to the term "numerical experimentation." There is no empirical content to these simulations in the sense that none of the inputs come from measurements or observations on real systems, and the lattice models are just that, mathematical models. It is the fact that they have a dynamical content due to being implemented on a real computational device that sets these models apart from ordinary mathematical models, because the solution process for generating the limit distribution is not one of abstract inference, but is available only by virtue of allowing the random walks themselves to generate the distribution. It thus occupies an intermediate position between physical experimentation and numerical mathematics, being identical with neither.

A quick comparison with the well-known Buffon's needle process is illuminating. Although this process is perhaps best known to philosophers for having induced Bertrand to invent his paradoxes of equiprobability, Buffon's original process is often considered to be the progenitor of empirical Monte Carlo methods. In its simplest form, the process consists of casting a needle of length L onto a set of parallel lines unit distance apart. (Here $L \leq 1$.) Then, by an elementary calculation, one can show that the probability that the needle will intersect one of the lines is equal to $2L/\pi$. As Laplace pointed out, frequency estimates of this probability produced by throwing a real needle can be used to empirically estimate the value of π, although as Gridgeman 1960 notes, this is a hopelessly inefficient method for estimating π, since if the needle is thrown once a second, day and night, for three years, the resulting frequency will estimate π with 95% confidence to only three decimal places. Be this as it may, the Buffon-Laplace method is clearly empirical in that actual frequencies from real experiments are used in the Monte Carlo estimator, whereas the standard methods for abstractly approximating the value of π using truncated trigonometric series are obviously purely mathematical in form. In contrast, the dynamic implementation of the Metropolis algorithm lies in between these two traditional methods. It is in this sense at least that I believe numerical experimentation warrants the label of a new kind of scientific method.

4.5

This point needs further discussion because it can easily be misconstrued. What of the fact that this is, after all, an *algorithm* within which even the random numbers are generated according to a deterministic formula? Doesn't this mean that in principle, this is no different from any traditional mathematical model and that our inability to carry out the requisite computations ourselves, and thus delegating them to an artificial computational device, is nothing but a practical concern and hence devoid of philosophical interest? Here the parallel with instrumentation is relevant. It is natural to respond to this fact by appealing to an "in principle" argument, and to use the contingency of our limited computational powers to argue that the inability to carry out such computations in practice is philosophically irrelevant, no matter how important it might be for the purpose of applied science. But here it is necessary to make a sharp distinction between on the one hand philosophical issues concerned with mathematics conceived as an autonomous discipline divorced from its scientific applications and on the other hand philosophical issues concerned with that subset of mathematics needed for, and sometimes developed in explicit recognition of, its applications in scientific models. One could conceive of this subset as merely being a part of pure mathematics, and that the philosophically important issues occur when considering the relation between the mathematics and reality. But that way of viewing computational science has the potential for transferring philosophical positions that are appropriate for the analysis of pure mathematics to the area of applied mathematics, where they are not so obviously appropriate. For us, the key concept is that of "computable." When philosophers use this term, they almost always have in mind the purely theoretical sense centered around Church's Thesis. But the dispositional aspect of the concept "computable" can also be construed as involving what can be computed in practice on existing computational devices. Consider now the parallel with the concept of "observable" mentioned in section 4.1. Many of the disputes within empiricism about the concept of observability hinged on whether it should be taken as fixed or as changeable. Those minimalist empiricists who took "observable" to be whatever is detectable by the unaided senses of humans rejected the idea that what was observable depended on upon the current state of technological enhancement of our sensory apparatus. For these minimalists, the epistemic acceptability of things beyond the immediately accessible was a matter of whether the terms referring to those perceptually inaccessible entities were reducible by definition to terms referring to things that were immediately accessible to the senses. In contrast for the nonminimalists, what was observable was (at least in part) a matter of what contingently available instrumentation was available.

The situation with the concept of "computable" is in some ways the inverse image of this situation regarding what is observable. Minimalist empiricists had a very narrow conception of what is observable, whereas the realm of what was viewed as observable by the technological nonminimalists was much wider. With the revised concept of "computable" the wide domain of recursive functions is drastically reduced in size to obtain the concept of computable that is appropriate for technologically accessible computational science. But within the domain of this technologically enhanced conception there remains a parallel, for what is considered computable goes far beyond what is actually calculable by the limited powers of a human.

4.6

It should be noted that there is a double element of contingency in this appeal to actual computability. It is obviously contingent upon what is currently available to us in terms of artificial computational devices, but it is also contingent upon the complexity of the world itself. For, if the number of degrees of freedom involved in physical systems was always very small, we should not need augmentation of our native computational abilities to employ the Monte Carlo methods described above. (We might need such augmentation to deal with other kinds of analytically intractable problems, but we are dealing here only with the case at hand. For some other cases where analytical intractability forces computational science on us, see Humphreys 1991.) But the world is not simple in that particular way and in order to apply our theories to systems with large number of degrees of freedom, such augmentation is forced upon us. This double element of contingency is not a reason to reject the position for which I have argued. In fact it is exactly the kind of consideration that *should* place constraints on an empiricist's epistemology, because what is acceptable to an empiricist has to be influenced by the contingencies of our epistemic situation rather than by appeal to superhuman epistemic agents free from the constraints to which we are subject. For in fact, an exactly parallel double contingency holds in the case of observability. How far the concept of "observable" gets extended beyond the realm of what is accessible to the unaided senses depends not only upon the current state of technological development in instrumentation, but upon what is in the world. If the world were composed entirely of relatively close, medium-sized objects devoid of microscopic internal structure, and it possessed only properties perceivable with our five senses, then we could not argue with the minimalist conception of observability. But of course the world is not that way, and so the minimalists do have to defend their limited conception of the observable.

I note in passing here that van Fraassen's recent (1980) attempt to return to a minimalist conception of what is observable is subject to objections along the above lines. For van Fraassen, the moons of Jupiter count as observable because we, as presently constituted humans, could see them directly from close up. But this possibility depends on technological advances just as much as does the observability (in the nonminimalist sense) of a charged elementary particle, which van Fraassen denies is observable. But why allow the use of technology in one case and not in the other? Even if it were said that the technological enhancement provided by spacecraft is an enhancement of a nonsensory faculty we have, that of locomotion, and is thus different in kind from the enhancement provided by bubble chambers, we could respond that our ability to observe something directly is a relation between us and the object, and it simply happens to be a linear spatial relation in the moons of Jupiter case, and a relative spatial size relation in the elementary particle case, if we are observing the particle, that is, rather than its charge. But these are not clearly relations of a different type.

So, there is indeed a significant parallel that can be drawn between the issues involved in instrumentational enhancement and those involved in computational science. The most immediate conclusion to draw is that in dealing with issues concerning the application of mathematical models to the world, as empiricists we should drop the orientation of an ideal agent who is completely free from practical computational constraints of any kind, but not restrict ourselves to a minimalist position where what is computable is always referred back to the computational competence of human agents. If, as many minimalist empiricists believe, it is impermissible to argue that humans could have had microscopes for eyes or could evolve into such creatures, and it is impermissible to so argue because empiricism must be concerned with the epistemological capacities of humans as they are presently constituted, then it ought to be equally impossible to argue that in principle humans could compute at rates 10^6 faster than they actually do. But that is just not plausible.

The position for which I have argued should be kept separate from an apparently similar issue that has been much discussed in the literature on connectionist architectures in artificial intelligence. One of the first applications of the Metropolis algorithm was to the process of simulated annealing, where a liquid that is cooled slowly ends up in an ordered state corresponding to an energy minimum. (In contrast, if the liquid is cooled quickly, it ends up in a state with a higher energy level.) The simulated annealing method has also been employed by connectionists, for example in Hopfield nets, to generate relaxation schedules for multiple constraint problems.[3] Because it is often asserted that one of the major differences between connectionism and classical (rule-based) artificial intelligence lies in the fact that the latter employs algorithms that are explicitly computable, whereas the former employs nonlinear functions that are not, and in consequence systems have to be allowed to settle themselves into the appropriate output state, it might seem that there is a similar lack of computational transparency in the Ising models. But this is not so. Although there are interesting problems posed by the use of continuous analog computational devices, the Monte Carlo methods described here are all computationally transparent, implementable on digital machines, and give finite approximations to the limit distribution.[4]

References

Binder, Kurt, and Dieter W. Heerman. 1988. *Monte Carlo Simulation in Statistical Physics.* Berlin: Springer-Verlag.

Gridgeman, Norman T. 1960. "Geometric Probability and the Number π." *Scripta Mathematica* 25: 183–95.

Humphreys, Paul. 1991. "Computer Simulations." In *PSA: Proceedings of the Biennial Meeting of the Philosophy of Science Association 1990*, vol. 2, edited by Arthur Fine,

---

[3] See, e.g., McClelland and Rumelhart 1986, chap. 7.

[4] I am grateful to Fritz Rohrlich for comments that helped improve this paper. The preliminary work for the paper was done under NSF grant # DIR 8911393, and an earlier version was read in June 1992 at the Venice conference on "Probability and Empiricism in the Work of Patrick Suppes."

Mickey Forbes, and Linda Wessels, 497–506. Chicago: University of Chicago Press for the Philosophy of Science Association.

McClelland, James L. and David E. Rumelhart. 1986. *Parallel Distributed Processing*. Vol. 1. Cambridge, MA: MIT Press.

Metropolis, Nicholas, Arianna W. Rosenbluth, Marshall N. Rosenbluth, August H. Teller, and Edward Teller. 1953. "Equation of State Calculations for Fast Computing Machines." *Journal of Chemical Physics* 21 (6): 1087–92.

Naylor, Thomas H. 1966. *Computer Simulation Techniques*. New York: John Wiley.

Rohrlich, Fritz. 1991. "Computer Simulations in the Physical Sciences." In *PSA: Proceedings of the Biennial Meeting of the Philosophy of Science Association 1990*, vol. 2, edited by Arthur Fine, Mickey Forbes, and Linda Wessels, 507–18. Chicago: University of Chicago Press for the Philosophy of Science Association.

van Fraassen, Bas. 1980. *The Scientific Image*. Oxford: Clarendon Press.

Templates, Opacity, and Simulations

Science has done many things for us, but one of its most spectacular achievements has been its ability to vastly expand our access to the world. In so doing, it has helped undermine the philosophical tradition of empiricism. Traditionally, empiricism restricted knowledge to what can be known through experience, and in the fifteenth century, what could be known through experience was a very small world indeed. It was a world without knowledge of ninth-magnitude stars, of bacteria or of viruses, of ultraviolet light, of 30 KHz acoustic waves, or of neon. Traditional empiricism cripples science by restricting its methods and data to anthropocentrically accessible sources, a position that is often motivated by the view that science is, contingently but inescapably, a set of activities carried out by humans. This is a view that is unsupportable for large areas of sophisticated modern science, and a more appropriate conclusion to draw from the science of the last fifty years is that the traditional focus on observables should be replaced by an emphasis on data, where data need not be directly accessible to humans. For scientific purposes, empiricism is dead.

A similar conclusion about mathematical methods runs through chapters 1 to 4 and the computational themes explored in Part I complement in certain ways some of the antireductionist themes of Part II. The importance of Philip Anderson's seminal article "More Is Different" (1972) lay in his arguments that it is often impossible to reconstruct the conceptual and theoretical apparatus of many areas of science from first principles. As a result, that apparatus must be taken on its own terms and results in a certain kind of conceptual emergence. In chemistry and elsewhere these difficulties are often computational in form, a fact that any account of theoretical reduction must take seriously.

Chapter 1 began life as a symposium presentation at the Philosophy of Science Association meetings in 1990.[1] Its central philosophical point is that applying a model or a theory to a concrete system requires far more than

---

[1] The physicist Fritz Rohrlich, another participant in the symposium, was enormously helpful to me both before and after the session. He knew a good deal more about computer simulations than I did, and if it had not been for his encouragement, I suspect that I would have abandoned the topic.

defining a mapping, postulating a morphism, or identifying similarity relations between a model and the world. Anyone who is seriously interested in how scientific theories and models are brought into contact with data must pay attention to the complicated process of moving from even modestly abstract models to the real world. This general moral is beautifully illustrated in Patrick Suppes's paper "Models of Data" (1962).[2]

Perhaps that moral is now more widely accepted that it was in 1990. During the discussion at the PSA session the view was often expressed that computer simulations were simply a matter of number crunching and devoid of any philosophical interest. Part of this attitude reflected the tendency of philosophers to dismiss calculation as intellectual sweatshop labor, a position at odds with the views of Johannes Kepler, John von Neumann, and Richard Feynman, all of whom were well aware of the importance of detailed calculations in applying science. This negative attitude toward calculations is reinforced by philosophy's natural and often justified tendency to abstract from practical details. The insistence that what is important are "in principle" methods has led to applied mathematics being treated with suspicion.[3] But applied mathematics is not just the application of pure mathematics. It has its own methods and culture that do not sit well with many standard views about mathematics in the philosophy of science and the philosophy of mathematics. For example, the kind of rigorous proof techniques with which we are all familiar in pure mathematics are often replaced in applied mathematics by heuristic procedures that are justified on inductive rather than deductive grounds. If you want to find the global minimum of a complicated function of three arguments, you often have to be satisfied with a trial-and-error technique such as the method of steepest descent. You are not guaranteed to find the absolute minimum value of the function with such a technique, but if the function is reasonably well behaved in a way that can be precisely specified, then you are likely after the application of the technique to arrive at a value that is close to that minimum.[4]

A similar trend in moving away from a reliance on in-principle definitions has been evident in computability theory. The original concepts of Turing computability, general recursive functions, Post computability, and so on have been supplemented by a finely grained set of degrees of computational complexity, the most famous of which is the P versus NP distinction. Some recent results concerning the undecidability or NP complexity of problems in physics show that there are limitations even in principle on the predictive abilities of

---

[2] For a discussion of Suppes's paper, see Humphreys 2015b, with a reply by Suppes 2015.

[3] Bill Wimsatt has been a long-term advocate of moving away from in-principle arguments. Numerous examples of this can be found in Wimsatt 2007.

[4] For a focused treatment of the role played by numerical methods in science, see Corless and Filion 2013.

scientific models. (See Istrail 2000; Barahona 1982; Gu et al. 2009; Humphreys 2015a.) Yet even those complexity results are theoretically oriented. The complexity classifications identify the worst-case scenario in a given class, and because most are still tied to the abstract concept of Turing computability, they remain detached from technological developments.[5]

Undoubtedly, many aspects of applying science in practice have little philosophical interest, being better addressed through science and technology studies. Addressing those topics that do have philosophical consequences, such as whether the representations used by machine-learning algorithms can be mapped onto representations that are understandable by humans, often requires guidance from results that have been developed within traditional philosophical domains that are remote from the application at hand. What is important is not to let existing philosophical attitudes unduly constrain either what is considered philosophically important or how one answers genuinely novel questions that arise within this new domain of philosophy of science. The epistemology of computational science is strikingly different from traditional epistemology.

Furthermore, in the computational domain, applied mathematics is discrete, essentially finitistic, and subject to serious technological limitations on what is feasible. Partly for those reasons it is a rich source of philosophical problems.

This is not the place to comment in detail on the many advances that have been made in our philosophical understanding of computer simulations since those early days, so I shall restrict my remarks to issues that were addressed in the original article, putting the points in a wider perspective where appropriate. For good overviews of the current state of the field see Lenhard 2016, Lenhard 2019, and Imbert 2017.

PI.1. Computer Simulations and Computational Science

Despite the title of chapter 1, the methodological shift that I had in mind is far larger than just computer simulations, which for understandable reasons have attracted the most attention from philosophers. The entire area of computational science, which includes computational physics, chemistry, biology, and other sciences, data analytics, optimization methods, visualizations, the production of simulated data, and even parts of the digital humanities, is rich with novel philosophical issues that have little to do with simulations per se. Each of these areas illustrates how much adjustment is needed in moving away from

---

[5] For an excellent survey of contemporary theories of computability, see Shagrir 2016.

the anthropocentric model of science within which the philosophy of science has long operated.

The degree to which the applicability of many scientific theories is both augmented and constrained by the techniques of computational science cannot be underestimated. Almost the entire field of contemporary computational chemistry would be impossible without the developments in hardware and in algorithms that have occurred since the 1950s.[6] The computational physicist Arno Schindlmayr has noted that a numerical approach to calculating the energies of an iron atom with twenty-six electrons requires 10^{78} function values and using a numerical approach on a solid would need $10^3 \times 10^{10^{23}}$ function values.[7] This fact nicely illustrates how even relatively simple atoms lie beyond the reach of ab initio methods, whether in physics or in other sciences. Although those results are relative to how the physical problem is represented, they are important pieces of evidence about the limits of scientific knowledge based on fundamental principles.

PI.2. Epistemic Opacity

Perhaps the most important epistemological consequence that computational science has for scientific methodology is epistemic opacity, the subject of section 2.4.2 of chapter 2 and section 3.3.1 of chapter 3. The definitions of epistemic opacity and essential epistemic opacity given there are relativized to an agent, but that is compatible with some processes being opaque or essentially opaque to an entire category of epistemic agents. Some examples are given below.

Epistemic opacity, although not under that name, played a role in some of the early discussions of computer-assisted proofs in mathematics (see Tymoczko 1979; Teller 1980), but those discussions revolved around whether the concept of proof had been altered by the introduction of computational methods. This was because parts of computer-assisted proofs are not surveyable, something that had previously been present in a different form for very long traditional proofs authored by multiple mathematicians. In those cases the epistemological problems are part of social epistemology, rather than individualistic epistemology, and are of a different kind than those introduced by computer-assisted proofs. One might, with considerable effort, combine the

---

[6] Hence the discussion of ab initio and semiempirical methods in computational chemistry in chapter 1. The importance of simulations in chemistry has been described in admirable detail by Lenhard 2014. Jeremiah James and Christian Joas 2015 have argued in great detail that in an earlier era applications in chemistry played a crucial role in the transition from the old to the new quantum theory.

[7] Lecture, ZiF (Bielefeld), June 6, 2013.

two with the aid of actor-network theory, an approach within which material objects can act as nodes in a social network.[8] Unsurveyable proofs are an example of essential epistemic opacity. It is not only that the programs are often too long to check in detail for every step and that the reliability of the compiler or translator must be known and that often the only way to solve the verification problem is to run the code on a concrete machine, but that the machine representations used in some computational tasks cannot be understood by humans, resulting in a machine-assisted epistemology.

What was equally important, and remains the central issue both in computer simulations and in computer-assisted proofs, is a loss of understanding as well as a change in the standards of justification. When analytic solutions or explicit proofs are available, we have a clear understanding of why the prediction, explanation, or conclusion follows from the basic principles of the theory, model, or axioms. It is this understanding that is often lacking in the use of simulations and that can lead to serious abuses of the method.

In the broader area of computational science, what is called explainable artificial intelligence (XAI) attempts to address the problem of epistemic opacity (see Lecun, Bengio, and Hinton 2015). This is not the place for an extended discussion of that enterprise, but one consideration is relevant to whether it can succeed. One area of interest for XAI is neural nets. If there is a significant similarity in computational architectures between those AI devices and human brains, and some version of the computational theory of mind is correct, then the following possibilities are relevant. Our cognitive representations are consciously accessible. We do not, qua ordinary humans, employ as part of our consciously accessible representations neural processes and states. So there is no reason to think that whatever representations are present at the neural level will be understandable by our conscious representations.[9] Perhaps some of them correspond to existing categories and concepts, such as edges, shadows, angles, and so forth, but some reasons would need to be given that they all do. To put it in somewhat anachronistic Kantian terms, an argument is needed to support the view that the categories used by specifically neural representations or by their machine analogues are understandable by the categorical apparatus of conscious human representations. Absent such an argument, XAI can only proceed by inductive success, not by a research program based on plausible presuppositions.

An analogy I used in *Extending Ourselves* (Humphreys 2004; see also Humphreys 2000) was between the use of scientific instruments to extend our perceptual capabilities and the use of computational devices to extend our mathematical abilities. Before discussing this analogy, a couple of remarks are

---

[8] See Latour 2005 for an introduction to actor network theory.
[9] The work on subconceptual and nonconceptual representations in Gunther 2003 is relevant here.

in order about the relation between this thesis and the well-known extended-mind thesis (Clark and Chalmers 1998). The similarity between the titles is purely coincidental and masks a fundamental difference of outlook.[10] Clark and Chalmers start with the human mind as the core cognitive agent and argue that some of the contents of that mind are located externally to the body that is associated with the mind in question. *Extending Ourselves* has a very different motivation; its primary aim is to show that the human mind is no longer the center of the epistemological universe and that as science progresses we should abandon it as the ultimate arbiter of justification. What is located externally to us are not the contents of minds but things that are significantly different, epistemically superior, and often inaccessible to human cognitive abilities.

Returning to the analogy between perceptual extensions and mathematical extensions, some aspects of the analogy are straightforward. Appeals to a priori mathematical intuitions are analogous to appeals to a posteriori perceptual intuitions, where perceptual intuitions are direct perceptual experiences. So the task for the mathematical realm that is analogous to the task for the perceptual realm is to show that there are methods for turning computer-generated mathematical results into a form that is directly accessible to humans. Formalism attempted a similar task, with some success. The difficulty of representing a problem in topology for which our topological imagination—one largely conditioned by our geometrical imagination—is inadequate can be overcome to some extent by a representation in a formal language. Perhaps automated proof checkers constitute a similar interface that allows us to have confidence in computer-generated proofs. By calibrating the process at key points to ensure conformity to traditional a priori results and then using warranted interpolations between those calibration points, a reliabilist epistemology can support a large number of conclusions based on computer-generated proofs.[11] The calibration points serve as the analogue in mathematics of overlap arguments that justify an appeal to scientific instruments in science.

Epistemic opacity leads to a version of the problem of scientific realism. Scientific realism as usually construed has to address the problem of how we can justify knowledge about unobservable entities that are accessible only through the use of scientific instruments. The problem of scientific realism is often taken to be tied to spatial scales; the realms of the very small and the very far away are central examples of how unobservability results from metrical constraints. Van Fraassen (1980, chap. 2) treats these two domains differently, entities in the realm of the very small remaining unobservable under

---

[10] A historiographic note: the paper "Extending Ourselves" (Humphreys 2000) originated as a talk of the same title that was given at the University of Pittsburgh in October 1997.

[11] For some suggestions about a related project see Lehrer 1995.

constructive empiricism, but things that would be observable were we to close the spatial gap are allowed to fall into the realm of the observable. For this position to be plausible, this hypothetical must tacitly appeal to the techno-logical availability of methods that allow us to put ourselves in a position to directly observe the object. Putting this much pressure on the dispositional content of "observable" leads us away from one of the central motivations for empiricism, which is the availability of methods for deciding epistemic disagreements. Was the dark side of the moon observable for Jules Verne in 1865 because he imagined traveling from the earth to the moon? And should the galaxy MACS0647-JD, which is 13.3 billion light years from Earth and is detectable only through the use of a gravitational lens in conjunction with the Hubble Space Telescope, be considered observable because in some counter-factual sense it could be observed directly?[12] Perhaps, but what is the epistemic value of that decision and why are some technologies such as spacecraft and cryogenic preservation acceptable in determining the domain of epistemic possibilities but others such as a simple test to detect the acidity of the soil in my back garden unacceptable?

Phenomena that are within the realm of the spatially observable can be-come unobservable because of inaccessibility in the temporal dimension. The motion of a hummingbird's wings in flight is a spatially macroscopic phe-nomenon but cannot be observed in ways required by either traditional or constructive empiricism. In order to do so, moderately high-speed cameras are required. To deny the evidence of the camera is to refrain from using the explanation that hummingbirds, observably winged at rest, use those wings to suspend themselves in midair. A philosophical possibility but an epistemo-logical peculiarity. In another dimension, sheer size can lead to ordinary mac-roscopic entities having unobservable properties, a fact that is now exploited by methods of computational textual analysis. Consider a project that has computationally analyzed over five million books in English. Even a relatively straightforward task such as identifying what percentage of word types that have been used in those books but that do not appear in standard reference dictionaries such as the *Oxford English Dictionary* and the *Merriam-Webster Unabridged Dictionary* is beyond the abilities of any single human. Using sam-pling techniques, the estimated answer is 52%, meaning that over half of the word types used in written English are not recorded in the most authoritative dictionaries (Michel et al. 2011, p. 177). This example also shows the asym-metry of the epistemology of discovery and the epistemology of confirma-tion. It is possible, given the list of words not in the dictionary, for a single

---

[12] I am assuming here that the observed redshifted light from the galaxy originated as visible, rather than ultraviolet, light. If it turns out that this is not true of that particular galaxy, one somewhat closer to Earth will serve as an example.

human to confirm the 52% figure by hand but not to have discovered the list unassisted. This type of example shows that the philosophical issues raised by computational methods are not restricted to the sciences but can appear in the humanities and everyday life. In the realm of computerized trading in financial markets, the flash crash of the New York Stock Exchange on May 6, 2010 has been reconstructed by the Federal Trade Commission, a process that took three months, and there are questions about whether the reconstruction is fully accurate. Because of this, there remains the central question of the extent to which humans are capable of fully understanding the details that constitute these massive computational manipulations of data.

Chapter 2 also draws a distinction between the hybrid scenario and the automated scenario. Although commentators have often expressed skepticism about the possibility of a fully automated science, it is frequently useful to think about epistemological problems from the perspective of a nonhuman agent. This is not easy, but at the very least the exercise immediately shows you that knowledge need not be connected with beliefs but can instead be grounded in more general kinds of representations. The idea of an automated science seems odd to many, but in the form of data analytics, automated data collection in astronomy, and automated genomic analysis in biology, it is already here. The results of the automated science still have to be made available to human scientists, and so we are still in the hybrid scenario, but we have arrived at point at which human intervention is minimal.

PI.3. Tractability and the Importance of Syntax

Another central claim in chapter 1 is the emphasis on the connection between the syntactic form of models and the tractability of the mathematics need to apply the model. In the paper, I make the case that apparently small changes in the syntactic form of the model can result in major changes in the difficulty of applying that model, a reason for paying close attention to the syntax of mathematically formulated theories. The emphasis on tractability runs counter to two traditions in the philosophy of science. The first is the emphasis on foundational studies. The second is the view that the primary task in applying a formal theory is to provide an appropriate semantics for the theory in order to connect the theory to the world. For many purposes, including representing abstract theory structure, those foundational approaches are entirely appropriate. They are less suited to capturing how theories and models are applied to real systems. A core example of this contrast is the long-running competition between linguistically oriented representations of theories (the syntactic view) and model-theoretic and state space representations (the semantic view). The semantic view is now dominant, but in virtue of abstracting from the linguistic particularities of the syntactic representation, it detaches itself from many of the methods that are crucial in bringing mathematically formulated theories to bear on natural systems. These methods are not, properly speaking, part of the theory itself, which is why abstracting from them is appropriate for foundational studies, but they are often heavily dependent for their success upon the specific syntactic form in which the theory is used.[13]

The insistence in section 1.3 of chapter 1 that a simulation must be computable in real time and implemented on a concrete machine is part of this move toward a more realistic account of how models are applied. Arguments that temporal logic can be used to represent the evolution of a simulation, and hence restore the appropriateness of in-principle approaches, miss the point in a very revealing way. For many computational processes, it is only by running the simulation on a concrete computer that we can verify the specific implementation of the algorithm. To accept this is not to adopt an antiphilosophical or nonphilosophical view. It is to recognize that there is an important epistemological difference between representation and application.

Many philosophers of science are deeply suspicious of introducing technological content into our subject, perhaps because the often ephemeral nature of particular technologies runs counter to an attachment to the view that philosophy is a generator of eternal truths, perhaps because some portion of the philosophy of technology was and is driven by political views. I sympathize with this resistance, but the boundaries between science, applied science, and technology are increasingly hard to identify. Progress in observational astronomy is primarily driven by progress in technology, whether in improved detector sensitivity, adaptive optics, or computational data processing. As one example, the Kepler satellite telescope selects which data to transmit to human observers on the basis of on-board algorithms. We humans never see the discarded data, or even know, except in a very general way, what they are. Although the algorithms for filtering the data are constructed by us, the possibility of observing unexpected anomalies in the unsent data has been severely reduced.

Bringing technology into consideration affects the answers we give to some philosophical questions. It is sometimes asked whether a particular scientific discovery, such as the discovery of the structure of DNA, could have occurred earlier than it actually did. Answers to that question are clarified when technological considerations are brought into play. The discovery of the structure of DNA crucially relied on the use of X-ray crystallography and would have been unlikely to have been confirmed without that or a similar imaging technology.

---

[13] Morrison 2015 emphasizes the point that methods such as variational techniques should not be considered as part of the theories to which they are applied. There is a discussion of the importance of tractability in Michael Redhead's seminal article "Models in Physics" (1980, sec. 5.iii) under the label "the computation gap," although the discussion is in terms of employing approximations rather than in reverting to computer-based methods.

PI.4. Definition of Simulations

A couple of comments on the working definition of simulations that was used in this paper are in order. A surprisingly large number of philosophers who have commented on this aspect of the paper have ignored the modifier "working."[14] This characterization of simulations was never intended as an explicit definition; rather, it was formulated in recognition of the fact that there is no generally accepted definition of a computer simulation and that to have introduced one would simply have invited unprofitable dickering over definitions. What a working definition is—and I hasten to add that this is not itself a definition—is a general description of the scope of the concept with no claim to have provided necessary or sufficient conditions. This allows room for disagreements over borderline cases. In later works I adopted a suggestion by Stephan Hartmann (1996) as a replacement for this working definition. Hartmann had noted that it is not uncommon for simulations to be carried out in cases where analytic solutions to the equations that constitute the model are available. That is correct. However, the original working definition has the virtue of representing an important epistemological difference between the cases where we are forced by necessity to use simulations because no analytic solutions are available and those cases in which simulations are used for other reasons. A superior definition of a computer simulation is given in Humphreys 2002 and in Humphreys 2004, 110–11.

In presenting the arguments in chapter 1, I was assuming that analytic solutions are generally preferable to numerical solutions and that one resorts to the latter only when the former are unavailable. In a recent paper, Vincent Ardourel and Julie Jebeile (2017) have provided a convincing set of reasons for adopting the converse position in certain cases. Rather than analytic solutions, they discuss analytical methods, where an analytical method is one that provides an exact solution, and their claim is that analytical methods for differential equations are often specific to that equation form, whereas numerical methods such as the collection of Runga-Kutta methods are more generally applicable. They also make the important point that in cases where the idealizations used in constructing a model are severe enough that model errors will be greater than numerical errors, concerns about the errors introduced by moving to a discrete model and the accompanying numerical imprecision due to truncation and other errors can sometimes be ignored when model errors are dominant.

---

[14] I also used a working definition of computational science in the "Numerical Experimentation" paper (chapter 4 of this volume), but this did not seem to arouse similar criticisms.

PI.5. Equation-Based Models and Agent-Based Models

One aspect of chapter 1 that in retrospect I would alter is its exclusive focus on equation-based models. Although that is a natural starting point for a discussion of computer simulations, the entire area of agent-based modeling is left unaddressed in the discussion. There are significant methodological differences between the two areas, but the main morals of this article carry over to other areas. This deficiency is partly remedied in chapter 2 and in chapter 15 together with the last postscript.

PI.6. Do Simulations Constitute a Revolution?

A recurring feature of the early philosophical discussions about computational science revolved around whether the introduction of computer simulation methods constituted a new way of doing science. Chapter 2 makes a case that they do. The opposite, critical, view holds that computational methods add nothing essentially new to the stock of scientific methods. Several decades on, the body of literature on computer simulations and related methods, together with numerous conferences and workshops on the topic, has made the latter position very difficult to defend.[15] A different question, that of whether computer simulations constitute a scientific revolution, is harder to decide.[16] In Thomas Kuhn's work, where some attention to practice plays a role, the emphasis on conceptual and theoretical change led to a diminution in the importance of methodological changes across scientific revolutions. In fact, the introduction of new methods can lead to what I call in chapter 2 an emplacement revolution, citing the advent of computational science as one example. As another example, the development of serious statistical methods in the late nineteenth and early twentieth centuries made possible modern sociology and econometrics, disciplines that in their current form could not have occurred prior to the invention of those techniques.

The concepts of a replacement revolution and an emplacement revolution that are introduced in chapter 2 are intended as supplements to the concepts of Kuhnian revolutions and Hacking revolutions. As I understand Kuhn—and multiple interpretations of his writings are possible—a Kuhnian revolution is always a replacement revolution since at least one of the four principal components of a paradigm has to be replaced in order

---

[15] The most prominent of these conferences is the Models and Simulations series of conferences, which is approaching its eighth iteration at the time of writing.

[16] Publications such as the NSF report *Report on Simulation-Based Engineering Science: Revolutionizing Engineering Science through Simulation* (National Science Foundation Blue Ribbon Panel 2006), constitute some evidence, but a principled set of considerations is needed in addition.

for a Kuhnian revolution to occur (see Kuhn 1970, postscript). An emplacement revolution is not the same as establishing a paradigm, although in rare
cases it could be if adding one of these four components is the final step
in constructing the paradigm. Perhaps Poincaré's early work on dynamical
systems theory allowed that field to become established. But in general the
two are different because by the definition of an emplacement revolution a
new method can be added to an existing paradigm while leaving in place the
methods, ontology, core representations, and exemplars that constitute the
original paradigm's identity conditions. With enough change, an emplacement revolution can turn into a replacement revolution. The development
of new methods in the form of data analytics to process enormous databases
might be seen as simply supplementing existing statistical methods that were
developed for smaller data sets, but Alvarado and Humphreys 2017 makes
a case that much of what goes on under the name of Big Data is genuinely
novel. If that is so, then a replacement revolution has taken place and a distinctively different field created. Sometimes the introduction of computational techniques into a field can be viewed by practitioners as incompatible
with analytic methods and hence, if widely adopted, would also lead to a
replacement revolution. Leombruni and Richiardi 2005 and Lehtinen and
Kuorikoski 2007 have argued that this negative attitude toward simulations
is present in mainstream Western economics and is thus an unnecessarily
conservative influence on economics.

Regarding Hacking revolutions, my reason for not using that concept was
primarily due to its lack of specificity, at least as concerns the case of computational science. More recently, in addressing Hacking's fourth condition,
Michael Mahoney (2011) has made a strong case in favor of the difficulties of
writing a complete history of computation. Schweber 2015 has effectively applied the idea of a Hacking revolution to quantum theory (including QFT), and
Schweber and Benportal 2015 have applied it to the quantum chemistry revolution, which includes computational chemistry. In light of these arguments,
I now recognize that the concept of a Hacking revolution can be useful, but the
different concept of an emplacement revolution adds a new dimension to the
framework of Kuhnian revolutions.

In a different way, computational templates have the potential to produce a
replacement revolution because, by emphasizing representations that are subject matter independent, they have the ability to replace scientific boundaries
that respect disciplinary domains by an organization of science that transcends
those boundaries, often quite radically. This has already happened in the area
of complexity theory, but the potential for reorganization is much greater. In
providing largely subject-matter-independent representations, computational
templates allow us to gain understanding across disciplinary boundaries.
This means that science is not as pluralistic an enterprise as has sometimes
been claimed. Unification can and is taking place not just through reductive

procedures but via cross-disciplinary representations that pick up on common structural features of different subject matters.[17]

PI.7. Logical Features of Simulations

Moving to addressing some logical features of the simulation relation, we can ask whether the simulation relation is symmetric. That is, if A is a simulation of B, is B always a simulation of A? The question is important because simulations are, among other things, representations of the system being simulated, and symmetry is often considered to be an objection to a proposed representation relation. My photograph represents me but not vice versa. Because A has, qua simulation, some representational or intentional component, whereas in most cases B, being a purely physical system, does not, in general, simulation relations are nonsymmetric. But if both A and B are simulations, and it is possible for one simulation to simulate another, then the simulation relation can be symmetric. A simulation and its target process do not have to run in synchrony, so two computers running simulations of one other could alternate their processing in a way that each simulated what was done during the last time step of the other. This would be a simulation in the sense that the physical operation of one machine, as well as the specific computations, were computationally modeled by the other. The argument for this uses the following distinction: system S provides a core simulation of an object or process B just in case S is a concrete computational device that produces, via a temporal process, solutions to a computational model that correctly represents B. If in addition the computational model used by S correctly represents the structure of the target system R, then S provides a core simulation of R with respect to B.[18] When system S_1 simulates the behavior of system S_2, which in turn simulates the behavior of system S_1, we have a case of symmetry. One way in which this can happen is when S_1 uses programming language L_1 to produce an output in programming language L_2 that when run on a concrete machine is a core simulation of R_2 with respect to B_2, i.e., a simulation of S_2. This in turn produces an output in L_1 that is a core simulation of R_1 with respect to B_1, i.e., a simulation of S_1. In that sense simulations can be different from representations. But there will be many cases in which the first simulation produces only a simulation of the behavior of the second, and that behavior is an insufficient basis for a simulation of the first system.

---

[17] For more explicit treatments of these topics, see Humphreys 2018; Knuuttila and Garcia-Deister 2018; and Herfeld and Döhne 2018.

[18] This is a slight modification of the definitions given in section 4.1 of Humphreys 2004.

In the case of the modeling relation, it is often required that if A is a model of B, then A must be simpler than B, which precludes symmetry. This requirement is usually imposed on abstract models and does not apply to concrete models, such as animal models of human physiology in which the model could be more complex than the target. In the case of computer simulations, there also is no need for a simulation to be simpler than the process it is simulating. For example, when what is being simulated is an experimental context for which many variables have already been controlled, digital approximations to continuous processes within the concrete experimental system will often introduce greater complexity into the simulation.

Is the simulation relation transitive? That is, if B is a simulation of A and C is a simulation of B, is C a simulation of A? In many cases transitivity will hold. But if it is a requirement for X to simulate Y that any approximations in X do not exceed a certain amount, then an accumulation of approximations could preclude long sequences of simulations of simulations from always passing on the property. Finally, is simulation reflexive? That is, does a process always simulate itself? It is certainly possible—consider a core simulation program in the above sense, say of planetary motion, running on a concrete machine, the output of which is a copy of the program itself that then feeds back into running again on that machine. The output from the simulation might be taken as a very indirect representation of the planetary motion and so could serve as a full representation of that behavior. However, most simulations will not be reflexive in this way.

PI.8. Numerical Experiments

The expression "numerical experiments" occurs in the original "Computer Simulations" article (chapter 1) as a term of art, and reflects its use in the literature on computer-assisted mathematics (Bowman, Sacks, and Chang 1993). There, "experiment" is used in contrast with "proof"; that is, if, after the numerical experiment has been conducted, no counterexamples are found to the given hypothesis, the hypothesis is asserted as correct with a very high degree of probability, rather than with certainty. Standard examples of this are probabilistic tests for primality of integers.[19] In a broader sense, what has come to be called experimental mathematics is a combination of trial-and-error methods, heuristics, and runs on real machines, sometimes with the aim of estimating a numerical value for a quantity that is difficult to arrive at by other means (see Borwein, Bailey, and Bailey 2004). This is the reason why I used the

---

[19] See, e.g., Rabin 1980. The test described therein will detect composite numbers with certainty but prime numbers only probabilistically.

expression "numerical experimentation" as the title of chapter 4. The Monte Carlo methods that are the main subject of the chapter fall into the category of numerical experiments qua estimation techniques, but the title has proved misleading despite my explicit denial that these methods are relevantly similar to those used in laboratory experimentation: "It [numerical experimentation] thus occupies an intermediate position between physical experimentation and numerical mathematics, being identical with neither" (1994, 112). The term "experiment" as used here and in experimental mathematics is a loose and informal sense of the term "experiment." There are generic similarities between numerical experiments and material experiments that consist in manipulation of variables, controlling for variables, and extrapolation from experimental results to a broader generalization, but these similarities do not provide an answer to the central philosophical question, which is: How, if at all, does the epistemological role of data produced from laboratory experiments differ from the epistemological role of data produced from computational methods?[20]

In Humphreys 2013 I addressed the issue of why running computations on concrete machines does not lead to the discovery of or generation of novel empirical facts. Conclusive reasons for maintaining a separation between computationally generated data and experimentally generated data have been given in Barberousse, Imbert, and Franceschelli 2009 and additional reasons can be found in Roush 2017. A philosophical issue that is related to their arguments is whether we must take into account the origins of the data in assessing that data. In the empiricist tradition, it was taken to be a great virtue of empirical data that they could be used independently of their origins.[21] This was an important epistemological advantage because if one ignored, or was ignorant of, the data's origins, the data could be analyzed and assessed on their own terms without the analysis being influenced by knowledge of those origins. In other traditions whether one can or should do that is much more controversial. When one is testing parametric statistical models, for example, the statistical models used often represent the conjectured structure of the system from which the data are collected, and model misspecification can lead to conclusions about the source of the data that are badly mistaken.

We can illustrate the role of data origins using a nonscientific example. The photographer Pavel Maria Smejkai has taken a series of images from war scenes, such as Joe Rosenthal's iconic photograph of the flag raising at Iwo Jima.[22] In that photograph Smejkai has digitally removed all of the human

---

[20] I restrict myself to a comparison with laboratory experiments. Field experiments, randomized controlled trials, natural experiments, and other forms of experimental techniques have their own peculiarities, but the central claims made here also apply to them.

[21] If the representation of empirical data inescapably contains theoretical elements, that theoretical content need not be about the system that generated the data.

[22] For an image see www.photoartcentrum.net/fatescapes05.html.

figures and the flag, leaving only an image of a landscape. The transformed Iwo Jima photograph has as its causal origins, hence what Barberousse et al. call data$_E$, a group of soldiers raising a flag. But after the transformations it is not about soldiers and a flag, even in part. It contains neither men nor a flag, and there are no other conventional referential devices that pick out the subject matter of the original photograph. The moral is that *computational transformations preserve causal origins but not referential content*.[23]

Some further clarification is required because the subject of chapter 4 involves the use of numerical experimentation methods within a model that has significant applications in physics. The case cited in "Numerical Experimentation" of the calculation of the value of π is a useful illustration of what is involved. One could approximate the value of π as the ratio of the measured circumference to the measured diameter of a real circle. As a sequence of real circles gives a closer approximation to a perfect circle, where a perfect circle is defined as a closed figure in which each point on the circumference is equidistant from an interior point, the sequence will convergence to a limit value. Alternatively one can construe the value of π as the limit of a trigonometric series or as the result of applying an iterative algorithm such as the Bailey-Borwein-Pflouffe spigot algorithm.[24] The Buffon needle experiment falls into the same class of physical experiments as does the process that involves measuring a sequence of physical circles because in each case the value of π can be estimated using an elementary mathematical argument applied to the data. (If the fact that π is defined as the ratio of the circumference to the diameter worries you, just measure the radius instead and double it.) The Buffon experiment can be computationally simulated, but simulating a physical process that calculates the value of a mathematical constant is different from simulating a physical process that calculates the value of a physical parameter, such as the probability of a die showing side 1. One reason for the difference is that the Buffon experiment requires a physical randomization process, such as spinning the needle with random initial conditions. If we wish to simulate such a process, then the relation between the physical randomization and the pseudorandom number generator must be described. One difference between the two is that we know the characteristics of the random number algorithm exactly, including the fact that it is periodic. In the physical randomizing process we have no such prior knowledge and must rely on statistical tests to justify the randomizing claim.

---

[23] It was possible to alter photographs in analog days, but that required a physical operation rather than a computational transformation.

[24] This algorithm, published in Bailey, Borwein, and Plouffe 1997, has the remarkable property of being able to calculate the nth digit of the decimal expansion of π directly. In this, it is unlike all previous algorithms for calculating π, in which each of the previous n − 1 digits must be calculated first.

These differences are pervasive in modeling. Consider a statistical model, which for simplicity we can take as a Bernoulli process. That is, the model consists of a stochastic process with discrete time steps, each iteration of the process is represented by a binomially distributed random variable (i.e., two possible outcomes, with probability p, $(1 - p)$ respectively, where p is constant across iterations), and the iterations are probabilistically independent. When using this model to represent a physical process, such as repeated tosses of a coin, we can justify its correctness or appropriateness in two ways. We can compare the data from tosses of a real coin, or data from an ensemble of replicas of that coin, with mathematical consequences of the model such as the variance of the distribution, and through the use of statistical estimators conclude for or against the adequacy of the model. Alternatively, one can try to directly verify that the assumptions of the model are satisfied by the real system. Even in this simple case, the second alternative is not entirely straightforward, for one has to infer from the supposition of causal independence of the tosses to their probabilistic independence. More generally, this second alternative always involves an inductive risk, which is that the system may be subject to influences of which one is ignorant because it is a naturally occurring system. The risk is evident in more complicated models applied to open systems, especially in the social sciences. For example, it is a standard assumption in linear causal models of social systems that the error terms are normally distributed, but it is in the nature of error terms that such assumptions cannot be directly verified.

One of the salient features of the models when computationally implemented is that we have direct access to the justificatory conditions because we built and implemented the model. This does not guarantee that the model's assumptions are indeed true in the computational implementation—this is a physical implementation with the attendant lack of an a priori certification—but our direct epistemic access to the model's assumptions does allow us to start with those assumptions, take the correct running of the model as paralleling the deductive consequences of the model, and in ordinary circumstances avoid the need to test the output data for conformity to the probability distribution connected with the model.

Finally, one topic that is mentioned only in passing in chapter 4 but deserves wider discussion is the status of closed-form solutions and representations. There is no standard definition of what counts as a closed-form solution, but one that would capture a considerable part of what are taken to be closed form solutions is this: a function f is a closed-form solution to a problem just in case f can be constructed from a bounded number of elementary operations on elementary functions. The elementary functions are characterized by a list. The list usually includes constant functions, addition, multiplication, division and root operations, elementary algebraic, and exponential and logarithmic functions and their inverses (Shanks 1993, 145; see also Chow 1999). Infinite series are not considered to be closed-form functions, and a well-known function that is not elementary is the normal probability distribution. It is possible that the appeal to elementary functions in the definition of closed-form solutions is a more subtle form of the anthropocentrism that is explored in the papers in Part I because the characterization of an elementary function is informal and is tailored to what is straightforwardly computable by humans.[25]

References

Alvarado, Rafael, and Paul Humphreys. 2017. "Big Data, Thick Mediation, and Representational Opacity." *New Literary History* 48 (4): 729–49.

Anderson, Philip W. 1972. "More Is Different." *Science* 177: 393–96.

Ardourel, Vincent, and Julie Jebeile. 2017. "On the Presumed Superiority of Analytical Solutions over Numerical Methods." *European Journal for Philosophy of Science* 7: 201–20.

Bailey, David, Peter Borwein, and Simon Plouffe. 1997. "On the Rapid Computation of Various Polylogarithimic Constants." *Mathematics of Computation* 66: 903–13.

Barahona, Francisco. 1982. "On the Computational Complexity of Ising Spin Glass Models." *Journal of Physics A: Mathematical and General* 15: 3241–53.

Barberousse, Anouk, Sara Franceschelli, and Cyrille Imbert. 2009. "Computer Simulations as Experiments." *Synthese* 169: 557–74.

Batterman, Robert W. 2007. "On the Specialness of Special Functions (the Nonrandom Effusions of the Divine Mathematician)." *British Journal for the Philosophy of Science* 58: 263–86.

Blue Ribbon Panel. 2006. "Revolutionizing Engineering Science through Simulation." *Report of the National Science Foundation Blue Ribbon Panel on Simulation-Based Engineering Science*, May. https://www.nsf.gov/pubs/reports/sbes_final_report.pdf

Borwein, Jonathan M., and David H. Bailey. 2004. *Mathematics by Experiment: Plausible Reasoning in the 21st Century*. Natick, MA: A.K. Peters.

Bowman, Kenneth P., Jerome Sacks, and Yue-Fang Chang. 1993. "Design and Analysis of Numerical Experiments." *Journal of the Atmospheric Sciences* 50: 1267–78. doi.org/10.1175/1520-0469(1993)050<1267:DAAONE>2.0.CO;2.

Chow, Timothy Y. 1999. "What Is a Closed-Form Number?" *American Mathematical Monthly* 106: 440–48.

Clark, Andy, and David Chalmers. 1998. "The Extended Mind." *Analysis* 58: 7–19.

Corless, Robert, and Nicholas Filion. 2013. *A Graduate Introduction to Numerical Methods: From the Viewpoint of Backward Error Analysis*. New York: Springer.

Gu, Mile, Christian Weedbrook, Alvaro Perales, and Michael Nielsen. 2009. "More Really Is Different." *Physica D* 238: 835–39.

Gunther, York H. 2003. *Essays on Nonconceptual Content*. Cambridge, MA: MIT Press.

---

[25] The definition of a Turing computable function is also one that seems to contain this tacit appeal to what is humanly computable. For a discussion see Shagrir 2016 and references therein. For arguments that the special functions have a more objective basis see Batterman 2007.

Hartmann, Stephan. 1996. "The World as a Process." In *Modelling and Simulation in the Social Sciences from the Philosophy of Science Point of View*, edited by R. Hegselmann, Ulrich Mueller, and Klaus G. Troitzsch, 77–100. Dordrecht: Kluwer Academic Publishers.

Herfeld, Catherine, and Malte Döhne. 2018. "The Diffusion of Scientific Innovations: A Role Typology." *Studies in the History and Philosophy of Science*. Forthcoming.

Humphreys, Paul. 1994. "Numerical Experimentation." In *Patrick Suppes: Scientific Philosopher*, vol. 2, *Philosophy of Physics, Theory Structure and Measurement Theory*, edited by Paul Humphreys, 103–18. Dordrecht: Kluwer Academic.

———. 2000. "Extending Ourselves." In *Science at Century's End: Philosophical Questions on the Progress and Limits of Science*, edited by Martin Carrier, Gerald Massey, and Laura Ruetsche, 13–32. Pittsburgh: University of Pittsburgh Press.

———. 2002. "Computational Models." *Philosophy of Science* 69: S1–S11.

———. 2004. *Extending Ourselves: Computational Science, Empiricism, and Scientific Method*. New York: Oxford University Press.

———. 2013. "What Are Data About?" In *Computer Simulations and the Changing Face of Experimentation*, edited by Eckhart Arnold and Juan Duran, 12–28. Cambridge: Cambridge Scholars Publishing.

———. 2015a. "More Is Different . . . Sometimes: Ising Models, Emergence, and Undecidability." In *Why More Is Different*, edited by Brigitte Falkenburg and Margaret Morrison, 137–52. Berlin: Springer.

———. 2015b. "Models of Data and Inverse Methods." In *Foundations and Methods from Mathematics to Neuroscience: Essays Inspired by Patrick Suppes*, edited by Colleen E. Crangle, Adolfo García de la Sienra, and Helen Longino, 61–68. Stanford, CA: CSLI Publications.

———. 2018. "Knowledge Transfer across Scientific Disciplines." *Studies in the History and Philosophy of Science*. Forthcoming.

Imbert, Cyrille. 2017. "Computer Simulations and Computational Models in Science." In *Springer Handbook of Model-Based Science*, edited by Lorenz Magnani and Tommaso Bertolotti, 735–81. Berlin: Springer.

Istrail, Sorin. 2000. "Statistical Mechanics, Three-Dimensionality, and NP-Completeness: 1. Universality of Intractability for the Partition Function of the Ising Model across Non-planar Surfaces." In *Proceedings of the Thirty Second Annual ACM Symposium on Theory of Computing*, 87–96. New York: ACM.

James, Jeremiah, and Christian Joas. 2015. "Subsequent and Subsidiary? Rethinking the Role of Applications in Establishing Quantum Mechanics." *Historical Studies in the Natural Sciences* 45: 641–702.

Knuuttila, Tarja, and Vivette Garcia-Deister. 2018. "Transfer and Templates in Scientific Modelling." *Studies in the History and Philosophy of Science*. Forthcoming

Kuhn, Thomas. 1970. *The Structure of Scientific Revolutions*. 2nd ed. Chicago: University of Chicago Press.

Latour, Bruno. 2005. *Reassembling the Social: An Introduction to Actor-Network-Theory*. Oxford: Oxford University Press.

LeCun, Yann, Yoshua Bengio, and Geoffrey Hinton. 2015: "Deep Learning." *Nature* 521 (7553): 436–44.

Lehrer, Keith. 1995. "Knowledge and the Trustworthiness of Instruments." *The Monist* 78: 156–70.

Lehtinen, Aki, and Jaakko Kuorikoski. 2007. "Computing the Perfect Model: Why Do Economists Shun Simulations?" *Philosophy of Science* 74: 304–29.

Leombruni, Roberto, and Matteo Richiardi. 2005. "Why Are Economists Sceptical about Agent-Based Simulations?" *Physica A* 355: 103–9.

Lenhard, Johannes. 2014. "Disciplines, Models, and Computers: The Path to Computational Quantum Chemistry." *Studies in History and Philosophy of Science Part A* 48: 89–96.

———. 2016. "Computer Simulations." In *The Oxford Handbook of Philosophy of Science*, edited by Paul Humphreys, 717–37. New York: Oxford University Press.

———. 2019. *Calculated Surprises*. New York: Oxford University Press.

Mahoney, Michael S. 2011. *Histories of Computing*. Cambridge, MA: Harvard University Press.

Michel, Jean-Baptiste, Yuan Kui Shen, Aviva Presser Aiden, Adrian Veres, Matthew K. Gray, the Google Books Team, Joseph P. Pickett, et al. 2011. "Quantitative Analysis of Culture Using Millions of Digitized Books." *Science* 331: 176–82.

Morrison, Margaret. 2015. *Reconstructing Reality*. New York: Oxford University Press.

National Science Foundation Blue Ribbon Panel. 2006. *Report on Simulation-Based Engineering Science: Revolutionizing Engineering Science through Simulation*. Washington, DC: NSF Press.

Rabin, Michael O. 1980. "Probabilistic Algorithm for Testing Primality." *Journal of Number Theory* 12: 128–38.

Redhead, Michael. 1980. "Models in Physics." *British Journal for the Philosophy of Science* 31: 145–63.

Roush, Sherrilyn. 2017. "The Epistemic Superiority of Experiment to Simulation." *Synthese*. Online first. doi.org/10.1007/s11229-017-1431-y.

Schweber, S. 2015. "Hacking the Quantum Revolution: 1925–1975." *European Physical Journal H* 40: 53–149.

Schweber, S. and G. Benporat. 2015. "Quantum Chemistry and the Quantum Revolution." In *Relocating the History of Science: Essays in Honor of Kostas Gavroglu*, edited by Theodore Arabatzis, Jürgen Renn, and Ana Simões, 41–66. Heidelberg: Springer.

Shagrir, Oron. 2016. "Advertisement for the Philosophy of the Computational Sciences." In *The Oxford Handbook of Philosophy of Science*, edited by Paul Humphreys, 15–42. New York: Oxford University Press.

Shanks, D. 1993. *Solved and Unsolved Problems in Number Theory*. 4th ed. New York: Chelsea Publishing Company.

Suppes, Patrick. 1962. "Models of Data." In *Logic, Philosophy, and Methodology of Science: Proceedings of the 1960 International Congress*, edited by Ernest Nagel, Patrick Suppes, and Alfred Tarski, 252–61 Stanford, CA: Stanford University Press.

———. 2015. "Reply to Paul Humphreys." In *Foundations and Methods from Mathematics to Neuroscience: Essays Inspired by Patrick Suppes*, edited by Colleen E. Crangle, Adolfo García de la Sienra, and Helen Longino, 253–55. Stanford, CA: CSLI Publications.

Teller, Paul. 1980. "Computer Proof." *Journal of Philosophy* 77: 797–803.

Tymoczko, Thomas. 1979. "The Four Color Theorem and Its Philosophical Significance." *Journal of Philosophy* 76: 57–83.

van Fraassen, Bas. 1980. *The Scientific Image*. Oxford: Clarendon Press.

Wimsatt, William C. 2007. *Re-engineering Philosophy for Limited Beings: Piecewise Approximations to Reality*. Cambridge, MA: Harvard University Press.

PART II

Emergence

{5}

How Properties Emerge

5.1. Introduction

Lurking in the shadows of contemporary philosophy of mind is an argument widely believed to produce serious problems for mental causation. This argument has various versions, but one particularly stark formulation is this:

(1) If an event x is causally sufficient for an event y, then no event x*
distinct from x is causally relevant to y (*exclusion*).

(2) For every physical event y, some physical event x is causally sufficient
for y (*physical determinism*).

(3) For every physical event x and mental event x*, x is distinct from x*
(*dualism*).

(4) So: for every physical event y, no mental event x* is causally relevant
to y (*epiphenomenalism*) (Yablo 1992, 247–48).[1]

This *exclusion argument,* as it is usually called, has devastating consequences for any position that considers mental properties to be real, including those nonreductive views that suppose mental properties to supervene upon physical properties. For if mental properties are causally impotent vis-à-vis physical properties, the traditional worry about epiphenomenalism confronts us: What is the point of having them in our ontology if they are idle? Abstract objects escape this worry, for we do not expect them to do causal work, but mental properties are retained in part because we believe them to affect the course of the world. If the exclusion argument is sound, then ratiocination, qualia, and the hopes and fears of mankind are simply smoke on the fire of brain processes.

---

[1] See Yablo 1992, 247 n. 5 for a partial list of versions of this argument that have appeared in the philosophical literature. I note here that he does not endorse the simple version of the exclusion argument because it is unsound.

This is bad enough, but there is a second argument, devised by Jaegwon Kim, that in conjunction with the exclusion argument seems to render nonreductive physicalism not merely uncomfortable but untenable, for it has as a conclusion that nonreductive physicalism is committed to the view that some mental properties must cause physical properties. Kim's argument, which I shall call *the downward causation argument*, was originally leveled against both nonreductive and emergentist approaches:

> But why are emergentism and nonreductive physicalism committed to downward causation, causation from the mental to the physical? Here is a brief argument that shows why. At this point we know that, on emergentism, mental properties must have novel causal powers. Now, these powers must manifest themselves by causing either physical properties or other mental properties. If the former, that already is downward causation. Assume then that mental property M causes another mental property M*. I shall show that this is possible only if M causes some physical property. Notice first that M* is an emergent; this means that M* is instantiated on a given occasion only because a certain physical property P*, its emergence base, is instantiated on that occasion. In view of M*'s emergent dependence on P*, then, what are we to think of its causal dependence on M? I believe that these two claims concerning why M* is present on this occasion must be reconciled, and that the only viable way of accomplishing it is to suppose that M caused M* by causing its emergence base P*. In general, the principle involved here is this: *the only way to cause an emergent property to be instantiated is by causing its emergence base property to be instantiated.* And this means that the "same-level" causation of an emergent property presupposes the downward causation of its emergent base. That briefly is why emergentism is committed to downward causation. I believe that this argument remains plausible when emergence is replaced by physical realization at appropriate places. (Kim 1992, 136)

Conjoin this second argument with the first, and you have more than mere trouble for nonreductive physicalism and emergentism, you have contradictory conclusions. (4) entails that there is no downward causation from the mental to the physical; the downward causation argument concludes that nonreductive physicalism and emergentism require such downward causation.

Something must go if mental properties are to survive, and that something is both arguments. Much is wrong with the exclusion argument, but what it shares with the downward causation argument is a pinched commitment to a dualist ontology, a laudable but usually unargued allegiance to the causal closure of the physical realm, and (nowadays) the idea that supervenience is the right way to represent the relation between the lower and higher levels of the world's ontology. Each of these is popular and each is wrong.

5.2. A Wider Perspective

The first thing to note about these arguments is that they are extremely general—they do not seem to rely on anything that is characteristic of mental properties, such as intentionality, lack of spatial location, having semantic content, and so on, and indeed it is often mentioned in passing that the arguments can be generalized to apply to a hierarchically ordered set of properties, each level of which is distinct from every other level.[2] If the exclusion argument does generalize to such hierarchies, and if, for example, chemical and biological events occupy higher levels than do physical events, then no chemical or biological event could ever causally influence a physical event, and if both arguments so generalize, then nonreductive physicalism leads to inconsistencies when applied to the general realm of the natural sciences too.

The situation is in fact more extreme than this, because most of our physical ontology lies above the most fundamental level, and in consequence only the most basic physical properties can be causally efficacious if these arguments are correct. Indeed, unless we have already isolated at least some of the most fundamental physical properties, every single one of our causal claims within contemporary physics is false and consequently there are at present no true physical explanations that are grounded in causes.

All of these are, of course, surprising and unwelcome consequences. We tend to believe that current elementary particle physics does reliably describe the formative influences on our world, that chemical corrosion can cause the physical failure of an aircraft wing, and that the growth of a tree can cause changes in the ground temperature beneath it. Yet our beliefs on this score may well be wrong. Perhaps the correct formulation of physicalism is a strict one: if, as many hold, everything is composed of elementary particles or fields, and the laws that govern those objects ultimately determine all else that happens, then there are fundamental physical events that are causally sufficient for aircraft wing failures, changes in soil temperatures, and all other physical events. This is a position that requires serious consideration, and indeed those versions of physicalism that require all nonphysical phenomena to supervene upon physical phenomena can be easily and naturally adapted to this kind of strict position. So we need to examine in detail a generalized version of each argument to see whether this radical physicalist conclusion can be supported and whether the problems for nonreductive physicalism extend all the way down to the penultimate level. There is a second reason for generalizing the arguments. It is preferable to examine arguments in a form in which the contextual assumptions are as transparent as possible. The relation between molecular chemistry and physics, say, is much more easily assessed than is the

---

[2] See, e.g., Yablo 1992, 247 n 5.

relation between physics and psychology because we have clearly articulated theories for the first pair, together with a reasonably clear set of constraints on the degree of reducibility of molecular chemistry to particle physics, but those relations are much murkier in the case of physics and psychology. In addition, by focusing on the lower levels of the hierarchy, we can avoid difficult problems involving the mental that are here irrelevant simply by recasting the arguments in a form that avoids reference to specifically mental properties.[3]

To generate the general arguments, we need a hierarchy of levels. For present purposes, I shall simply accept that this can be done in whatever way the reader finds most congenial. (The strata must correspond to some real differences in ontological levels rather than to a mere set of epistemic distinctions.) For concreteness, we can think in terms of this assumption: (L) There is a hierarchy of levels of properties $L_0, L_1, \ldots, L_n \ldots$ of which at least one distinct level is associated with the subject matter of each special science, and L_j cannot be reduced to L_i for any $i < j$.

I shall not try here to give any additional criteria for distinguishing levels. Rather, I am simply adopting, for the purposes of the argument, the abstract assumption of both arguments that there is indeed some such hierarchy. At the very least, one would have to consider this an idealization of some kind, and we shall see that the assumption that there is a discrete hierarchy of levels is seriously misleading and probably false. It seems more likely that even if the ordering on the complexity of structures ranging from those of elementary physics to those of astrophysics and neurophysiology is discrete, the interactions between such structures will be so entangled that any separation into levels will be quite arbitrary.

5.3. The Generalized Exclusion Argument

As a general principle, premise (1) is false as we originally formulated it, for it asserts that, first, causal antecedents of x and, second, events causally intermediate between x and y are causally irrelevant to y, and this is obviously wrong.

So, let us use the preliminary definition:

An event z is *causally connected* to a second event x if and only if x causes z or z causes x; z is *causally disconnected* from x just in case z is not causally connected to x.

---

[3] This allows, of course, that the realm of the mental does generate additional peculiar problems. The exclusion and downward causation arguments are entirely independent of those peculiarities, however, except perhaps for a prejudice against the mental. Were it not for that prejudice, the exclusion argument could be run in reverse to exclude physical causes, a feature emphasized to me by Peter Dlugos.

The proper formulation of (1) is then:

(1′) If an event x is causally sufficient for an event y, then no event x*
distinct from x and causally disconnected from x is causally relevant
to y. (exclusion)[4]

To make the revised principle true requires a strict criterion of event identity so that in particular, the exact time and way in which an event occurs is crucial to that event having the identity it does. This criterion is needed to exclude cases where x is sufficient for y but x*, which is causally disconnected from x, brings about (a somewhat earlier analogue of) y before the connecting process from x has brought about y.

Such a strict criterion, called "fragility" by David Lewis,[5] is controversial, and rightly so, but I shall accept it here simply because its adoption avoids distracting issues and does not affect the essential features of the argument.

Suppose now that we accept principle (L) and ask whether an analogue of premise (2) is plausible for each level L. What I shall call i-determinism (which is what a generalization of (2) asserts) is different from i-closure, the thesis that all events that are causally relevant to a given i-level event are themselves i-level events: i-determinism is compatible both with upward and with downward causation, where an i-level event causes another i-level event through a chain involving events at other levels. To see that (2) is not true when generalized, consider (2) with "physical" replaced with "biological." It might now be true that for every future biological event there is some biological antecedent that guarantees it. But to assert biological determinism for every biological event in the history of the universe would be to rule out what is commonly believed, which is that biological phenomena were not always present during the development of the universe.[6] The first biological event, under whatever criterion of "biological" you subscribe to, must have had a nonbiological cause. So a generalization of (2) is not plausible for any level above the most fundamental level of all, which we call the 0-level, and so we shall restrict ourselves to a formulation of (2) for that level only, the level of whatever constitutes the most fundamental physical properties, viz.:

(2′) For every 0-level event y, some 0-level event x is causally sufficient for
y. (0-level determinism)[7]

---

[4] I am here setting aside overdetermining events as genuine examples of causation. Although some discussions of the exclusion argument see acceptance of overdetermination as a way out of the difficulty, this is not a convincing move, and I shall not follow that route. Not the least reason for this is that cases of simultaneous overdetermination are exceedingly rare.

[5] See Lewis 1986, postscript E. Lewis rejects extreme versions of the fragility approach.

[6] This temporal development gives rise to evolutionary emergence. I shall not pursue that topic here but the reader can easily develop such an account from materials in this paper.

[7] Premise (2′) is, on current evidence, false because fundamental physics appears to be indeterministic in certain respects. I have preserved the original form of the argument to keep things simple, but

The third premise requires a criterion of distinctness of events (as does (1′)). This cannot be done in terms of spatiotemporal distinctness alone, because on supervenience accounts the supervening event is spatiotemporally coincident with the subvenient event(s). So the burden of characterizing distinctiveness will have to lie on principle (L) and we shall take as a sufficient condition for two events being distinct that they occupy different levels in the hierarchy. With this understanding we have:

(3′) For every 0-level event x and every i-level event x_i^* $(i > 0)$ x is distinct from x_i^*, (pluralism)

Then it follows immediately that

(4′) For every 0-level event y, no i-level event x_i^* $(i > 0)$ that is causally disconnected from every 0-level event antecedent to y is causally relevant to y.

This modified conclusion shows how nonreductive physicalism can avoid the conclusion of the simple version of the exclusion argument and hence can also avoid the overall contradiction with the conclusion of the second argument, because the conclusion (4′) allows higher-level events to causally affect 0-level events if the former are part of causal chains that begin and end at the 0-level.[8]

In order for us to use that possibility to argue for emergent properties, we need to address the second argument, which among other things, precludes higher-level causal chains that do not involve 0-level events.

5.4. Generalizations of the Downward Causation Argument

In looking at generalizations of the downward causation argument, it is worthwhile to again lay out explicitly the assumptions which underlie it. The first is a supervenience assumption that permeates the contemporary literature on nonreductive physicalism and is retained by Kim for emergent properties.[9]

---

to allay worries about the truth status of (2′), one can reformulate the argument thusly: Replace "causally sufficient" in (1′) and (2′) by "causally complete," "event" by "set of events," and x by $\{x_i\}$, where "causally complete" means either that all events necessary in the circumstances for y are included in the set or that all events that are probabilistically relevant to y are in the set. See, e.g., Lewis 1986 for one account of the former, Humphreys 1989 for one account of the latter. (3), (4) remain as they are. (2′) will be false if the universe had a first uncaused event, but that fact is irrelevant here.

[8] More complex versions of the argument obviously allow similar possibilities for causal chains beginning and ending at higher levels than 0.

[9] Kim actually examines both the realizability and the supervenience approaches. I restrict myself to the latter.

(5) Every emergent property is supervenient upon some set of physical properties.

One natural generalization of this is

(5') Every j-level property $(j > 0)$ is supervenient upon some set of i-level properties, for $i < j$.

Strict physicalists might want to insist that all higher-level properties supervene upon 0-level properties. However, unlike the first argument, where premise (2) was true only for 0-level properties, here we can maintain full generality. In fact, if there are emergent properties, the strict physicalist position will be false, and we shall have to leave room for some (nonemergent) properties to supervene upon j-level emergent properties but not upon 0-level properties alone. Next, the assumption explicitly cited by Kim:

(6) The only way to cause an emergent property to be instantiated is by causing its (set of) emergence base properties to be instantiated.

Its generalization will be (assuming that supervenience is a transitive relation):

(6') The only way to cause a j-level property to be instantiated is by causing a set of i-level properties $(i < j)$, the subvenient basis, to be instantiated.

Premises (5') and (6') are closely related, for the plausibility of (6') rests on accepting something like (5'), the idea being that emergent properties cannot exist separately from whatever physical properties give rise to them.

Then we have the important condition

(7) A property is emergent only if it has novel causal powers[10]

We can retain this unchanged for the generalized argument.

5.5. An Emergentist Answer to the Second Argument

How can we now escape the conclusion of the downward causation argument? We can begin by refining the event ontology used in the argument, which appeals to properties as causes. This way of speaking, about property causation, is clearly an abbreviation for an instance of one property causing

---

[10] In fact, as Martin Jones pointed out to me (pers. comm.), the novelty of the causal powers seems to play no role in Kim's central argument. Even if mental properties produced familiar physical consequences that could also be brought about by physical properties, the argument would still hold. The use of novelty is primarily in characterizing the difference between emergent and nonemergent properties, for it is an essential feature of emergent properties that they be new.

an instance of another property. We shall have to resort here to a certain amount of notational clutter. This will not be pretty, but it has a certain suggestiveness that may be helpful. From here on I shall talk of i-level properties and entities, P_m^i and x_r^i, respectively, for $i \geq 0$. I call a property (entity) an i-level property (entity) if i is the first level at which instances of P_m^i (x_r^i) occur. These i-level properties may, of course, also have instances at higher levels as, for example, physical properties such as mass and volume do. We need to keep distinct i-level entities and i-level properties, for it is possible that, in general, i-level entities may possess j-level properties, for $i \neq j$. However, for simplicity in what follows we shall assume that i-level properties are instantiated by i-level entities. So we have that $P_m^i(x_r^i)(t_1)$ causes $P_n^i(x_s^i)(t_2)$ where $t_2 > t_1$ and x_r^i, x_s^i the entities possessing the properties, may or may not be the same.[11] Here P_m^i is the mth i-level property; x_r^i is the rth i-level entity, and $P_m^i(x_r^i)(t_k)$ denotes the instantiation of P_m^i, by x_r^i at time t_k.

Suppose now that the i-level properties constitute a set $I = P_1^i,...,P_n^i ...$[12] and that these i-level properties are complete in the sense that I is exhaustive of all the i-level properties. Now introduce a *fusion operation* [.*.], such that if $P_m^i(x_r^i)(t_1)$, $P_n^i(x_s^i)(t_1)$ are i-level property instances, then $P_m^i(x_r^i)(t_1) * P_n^i(x_s^i)(t_1)$ is an i + 1-level property instance, the result of fusing $P_m^i(x_r^i)(t_1)$ and $P_n^i(x_s^i)(t_1)$. I want to emphasize here that it is the fusion operation on the property instances that has the real importance for emergence. Usually, the fusion operation acting on objects will merely result in a simple concatenation of the objects, here represented by $(x_r^i) + (x_s^i)$, within which the individuals x_r^i and x_s^i retain their identities. However, as we noted earlier, for full generality we would need to allow for the possibility of new i + 1-level objects.[13] Moreover, fusion usually is not instantaneous, and we should represent that fact. So I shall represent the action of fusion by $P_m^i[(x_r^i)(t_1) * P_n^i(x_s^i)](t_1) = P_m^i * P_n^i[(x_r^i) + (x_s^i)](t_1')$.[14] For simplicity, I shall assume here that * itself is an i-level operation (i.e., that it is an operation of the same level as the property instances which it fuses).

---

[11] Kim asserts (1992, 123) that the emergentists of the 1920s held that no new entities emerged at new levels of the hierarchy, only new properties. Allowing that to be historically accurate, it is as well to allow, at least notationally, that we might have new entities emerging as well as properties.

[12] The cardinality of this set is unrestricted—the integer subscripts are used for convenience only.

[13] Supervenience advocates have also recognized the need for this. See, e.g., Kim 1988.

[14] There is a notationally harmless but metaphysically important ambiguity here between fusion operations on property instances and on properties. The latter is metaphysically derivative from the former in that when $P_m^i\left(x_r^i\right)\left(t_1\right) * P_n^i\left(x_s^i\right)\left(t_1\right)$ exists, then there is by virtue of this an instance of a novel property, signified by $P_m * P_n$, at level i + 1. This disambiguation sits most happily with the position that fusion brings into being new properties, a position that seems to fit well with the idea of emergence. Those who subscribe to the view that there are eternal emergent properties that are uninstantiated prior to some time can think of $P_m * P_n$ as a mere notational device indicating a move to a previously uninstantiated property at a higher level.

By a fusion operation, I mean a real physical operation, and not a mathematical or logical operation on predicative representations of properties. That is, $*$ is neither a logical operation such as conjunction or disjunction nor a mathematical operation such as set formation.[15] $*$ need not be a causal interaction, for it can represent interactions of quite different kinds.

The key feature of $[P_m^i * P_n^i][(x_r^i)+(x_s^i)](t_1')$ is that it is a unified whole in the sense that its causal effects cannot be correctly represented in terms of the separate causal effects of $P_m^i(x_r^i)(t_1)$ and of $P_n^i(x_r^i)(t_1)$. Moreover, within the fusion $[P_m^i * P_n^i][(x_r^i)+(x_s^i)](t_1')$ the original property instances $P_m^i(x_r^i)(t_1)$, $P_n^i(x_s^i)(t_1)$ no longer exist as separate entities and they do not have all of their i-level causal powers available for use at the i + 1st level.[16] Some of them, so to speak, have been "used up" in forming the fused property instance. Hence, these i-level property instances no longer have an independent existence within the fusion. In the course of fusing they become the i + 1-level property instance, rather than realizing the i + 1-level property in the way that supervenience theorists allow the subvenient property instances to continue to exist at the same time as the supervenient property instance. For example, the cusped paneling and brattishing that makes the fan vaulting of the King's College Chapel architecturally transcendent exists simultaneously with the supervenient aesthetic glories of that ceiling. In contrast, when emergence occurs, the lower-level property instances go out of existence in producing the higher-level emergent instances. This is why supervenience approaches have great difficulty in properly representing emergent effects. To see this, consider the following formulation of strong supervenience:[17]

A strongly supervenes upon B just in case, necessarily, for each x and each property F in A, if x has F, then there is a property G in B such that x has G and necessarily if any y has G, it has F. (Kim 1984, 165)

Now let A be the fusion $[P_m^i * P_n^i][(x_r^i)+(x_s^i)](t_1')$. Upon what can this supervene? Because we are here considering only the abstract possibility of emergent features, consider a very simple world in which $P_m^i(x_r^i)(t_1)$ and $P_n^i(x_r^i)(t_1)$ are the only i-level property instances occurring at t_1, and in which there are no i-level property instances at t_1'. Then, trivially, there is nothing at t_1' at the i-level upon which $[P_m^i * P_n^i][(x_r^i)+(x_s^i)](t_1')$ can supervene. Faced with this, the supervenience advocate could try a different strategy, one that relies on

---

[15] In contrast, it is standard in the literature on supervenience to construe the subvenient basis in terms of sets of properties, or in terms of a disjunctive normal form of properties, where it is assumed that it makes sense to perform logical operations on properties. These devices are inappropriate for characterizing emergent properties and are a legacy of the continuing but, in my view, fruitless attempt to reconstruct causation and associated concepts logically rather than ontologically.

[16] As mentioned earlier, the objects themselves will often retain their separate identities.

[17] This argument carries over, with simple modifications, to the definition of weak supervenience.

the fact that the definition of strong supervenience does not require the supervenient and subvenient instances to be simultaneous. So one could use the earlier instances $P_m^i(x_r^i)(t_1)$ and $P_n^i(x_r^i)(t_1)$ themselves as the base upon which the later $[P_m^i * P_n^i][(x_r^i) + (x_s^i)](t_1')$ supervenes. Yet once one has allowed this temporal gap, the supervenience relation is in danger of collapsing into an ordinary causal relation. In order for the base instances to (nomologically) necessitate the fusion instance, the absence of all intervening defeaters will have to be included in the subvenient base, and this will give us a base that looks very much like a Millian unconditional cause.[18] Whatever the supervenience relation might be, the way it is used in nonreductive physicalism is surely not as a causal relation, because that would immediately convert nonreductive physicalism into old-fashioned epiphenomenalism.

A second reason why supervenience seems to be an inappropriate representation of certain cases of fusion is given by the physical examples in section 5.6, and I thus refer the reader to that section. However, because my purpose here is not to attack supervenience, but rather to provide a solution to the problem of upward and downward causation, I now return to that issue.

It has to be said that the unity imposed by fusion might be an illusion produced in all apparent cases by an epistemic deficit, and that when properly represented all chemical properties, for example, might be representable in terms of the separable (causal) properties of their chemical or physical constituents. But whether this can be done or not is, of course, the issue around which emergentism revolves and I shall address it explicitly in a moment in terms of some examples. What I maintain here is this: that one comprehensible version of emergentism asserts that at least some i + 1-level property instances exist, that they are formed by fusion operations from i-level property instances, and that the i + 1-level property instances are not supervenient upon the i-level property instances. With this representation of the emergent i + 1-level property, let us add a claim that is characteristic of many versions of nonreductive physicalism, especially those motivated by multiple realizability considerations: the token identity of i + 1-level property instances and fusions of i-level property instances. That is, although we cannot identify the property P_l^{i+1} with $P_m^i * P_n^i$ when P_l^{i+1} is multiply realizable, we *can* identify some instances of P_l^{i+1} with some instances of $P_m^i * P_n^i$. It is important to remember for the purposes of the present argument that we are concerned with causal and other interactions, and not with the problem of how properties themselves are related across levels. The latter is the focus of greatest concern within nonreductive physicalism, but our problems can be solved without addressing the interlevel relationships of the properties themselves. In fact, given that the higher-level properties are emergent, there is no reason to identify them with,

---

[18] See Humphreys 1989, sec. 25, for a discussion of this condition.

or to reduce them to, combinations of lower-level properties. Coupled with our previous reminder that it is property instances that are involved in causal relations, and not properties directly, we now have a solution to the downward causation problem for the case of emergent properties.[19]

Suppose that $P_l^{i+1}(x_l^{i+1})(t_1')$ causes $P_k^{i+1}(x_k^{i+1})(t_2')$, where both of these instances are at the $i + 1$-level. What we have is that the i-level property instances $P_m^i(x_r^i)(t_1)$ and $P_n^i(x_s^i)(t_1)$ fuse to produce the $i + 1$-level property instance $[P_m^i * P_n^i][(x_r^i)+(x_s^i)](t_1')$, which is identical with $P_l^{i+1}(x_l^{i+1})(t_1')$. This $i + 1$-level property instance then causes the second $i + 1$-level property instance $P_k^{i+1}(x_k^{i+1})(t_2')$. This second $i + 1$-level property instance, *if it is also emergent*, will be identical with, although not necessarily result from, a fusion of i-level property instances $[P_r^i * P_s^i][(x_u^i)+(x_v^i)](t_2')$. But there is no direct causal link from the individual property instances $P_m^i(x_r^i, t_1)$ and $P_n^i(x_s^i, t_1)$ to the individual decomposed property instances $P_r^i(x_u^i)(t_3)$ and $P_s^i(x_v^i)(t_3)$.

Diagrammatically, we then have

$$P_l^{i+1}(x_l^{i+1}, t_1') \quad\quad —\text{causes}—> \quad\quad P_k^{i+1}(x_k^{i+1}, t_2')$$

$$\text{(is identical with)} \quad\quad\quad\quad\quad\quad \text{(is identical with)}$$

$$[P_m^i * P_n^i][(x_r^i)+(x_s^i)](t_1') \quad\quad\quad\quad [P_r^i * P_s^i][(x_u^i)+(x_v^i)](t_2')$$

$$\nearrow <—> \nwarrow \quad\quad\quad\quad\quad\quad\quad \swarrow <—> \searrow$$

$$\text{(fuses)} \quad\quad\quad\quad\quad\quad\quad\quad\quad \text{(decomposes)}$$

$$P_m^i(x_r^i)(t_1) \quad P_n^i(x_s^i)(t_1) \quad\quad\quad P_r^i(x_u^i)(t_3) \quad P_s^i(x_v^i, t_3)$$

I note here that decomposition does not have to occur. The system might stay at the $i + 1$st level while it produces further $i + 1$-level effects. Nor is it necessary that P_k^{i+1} be an emergent property. When it is not, the identity on the right side of the diagram will not hold, and the lower two tiers on the right will be missing. Perhaps some $i + 1$ property instances are primitive in this way, but this is doubtful given what we know of the evolution of our universe.

One further source of concern exists and it relates directly to assumption (6′) of the downward causation argument. Is it possible for $i + 1$-level instances to directly produce other $i + 1$-level instances without synthesizing them from

---

[19] I want to emphasize here that what follows is to be construed only as a representation of the correct relationship between emergent property instances and property instances on lower levels when causal sequences are involved. It is not to be construed as an argument that in all cases where we have different levels of property instances, emergentism holds. Supervenience does have some restricted uses.

lower-level instances? These higher-level instances are usually emergent, and so it might be thought that they must themselves be formed by fusion from lower-level instances and not by direct action at the higher level. This concern fails to give sufficient credit to the ontological autonomy of emergent property instances. Recall that i-level instances no longer exist within i + 1-level instances—the higher-level instances act as property instance atoms even though they may, under the right circumstances, be decomposed into lower-level instances. It is perfectly possible for an i + 1-level instance to be directly transformed into a different i + 1-level instance (often with the aid of other property instances) or to directly transform another, already existing, i + 1-level property instance (again usually with the aid of other property instances). Simply because the i-level instances no longer exist, they can play no role in this causal transformation.

We can now see what is wrong with premises (5') and (6'). (5') misrepresents the way in which emergent property instances are produced, for as we have seen, the relationship between the higher and lower levels is not one of supervenience. A last reply is available to supervenience advocates insisting on the need for relations between properties. If the emergent instance is produced by a causal interaction, they can insist that causes require laws and that the generality inherent in laws requires properties as well as instances. This is not something that a thoroughgoing ontological approach needs to accept. Singular causes can be taken as fundamental,[20] and whether or not causal laws (or their statements) can then be formulated in terms of relations between properties (predicates) depends upon the complexity of the part of the world involved. Sometimes they can be, often they cannot.

We have already seen the problem with (6'): it is false to say that the i-level property instances co-occur with the i + 1st-level property instance. The former no longer exist when they fuse to form the latter. That is why the notation used here is not entirely adequate, but this is almost inescapable given that formal syntax is ordinarily compositional. $[P_r^i * P_s^i][(x_u^i) + (x_v^i)](t_2')$ is not "composed of" $P_r^i(x_u^i)(t_3)$ and $P_s^i(x_v^i)(t_3)$: a physical process of decomposition (better "defusion") is required to create the last two. So, we have given a construal of what i + 1-level emergent properties might be, shown how they can have causal properties that are new in the sense that they are not possessed by their i-level origins, and provided the appropriate sense in which the emergent properties are instantiated only because lower-level properties are instantiated. But there is now no sense in which, as the exclusion argument claims, i-level property instances are overdetermined by a combination of previous i-level instances and i + 1-level instances. Nor are the i + 1-level instances always required to be brought into being by some simultaneous set of i-level property instances, as the downward causation argument asserts. I believe that there are two reasons

---

[20] See Humphreys 1989, sec. 25.

why a solution of this kind is not immediately obvious when reading these arguments. First, by representing the emergence base within the downward causation argument as a single, unanalyzed property P*, the fact that emergent property instances are formed by a fusion process on lower-level instances cannot be accurately represented. Second, by representing the situation in terms of properties rather than property instances, supervenience seems to be forced upon us, when in fact a treatment purely in terms of instances is open to us.

We do, of course, lose the causal closure of i-level property instances. This is not something that should disturb us overmuch. If a picture like the one I have described is roughly correct, then it turns out that an undifferentiated commitment to "physicalism" is too crude and that both the "mental" and the "physical" are made up of multilayered sets of strata, each level of which is emergent from and (probably) only arbitrarily separable from the layers beneath it.

To sum up what we have established: the claim that an i + 1-level emergent property is instantiated only because its i-level emergence base is instantiated is wrong—the "emergence base" is not the reason the emergent property is instantiated—it is the move from the i-level to the i + 1-level by fusion that gives us emergence. Second, the problem of overdetermination that is lurking behind downward causation is now less problematical. We may maintain that all i-level events are determined by i-level antecedents but often this will be by way of j level intermediaries.

So, to conclude: we have seen that two intriguing and widely canvassed arguments against emergent properties do not succeed in establishing their conclusions. It is more important, though, to emphasize that a robustly ontic attitude toward emergent properties, rather than the more common logical approaches, can give us a sense of what emergent features might be like. Most important of all, I think, is to stop thinking of these issues exclusively in terms of mental properties, and to look for examples in more basic sciences.

In addition, construing the arguments in terms of multiple layers of property instances reminds us of two things: that there are many, many levels of properties between the most fundamental physical level and the psychological, and that it is left unacceptably vague in much of the philosophical literature just what is meant by "physical." Being reminded of the large variety of property levels below the psychological, some of which are arguably emergent, should at least make us aware of the need to be more explicit on that score, and that *some* of the mysteries surrounding the physical/mental cleavage are perhaps the result of an inappropriate dichotomy.

5.6. From Metaphysics to Physics and Back

Thus far, the discussion has been completely abstract. My intention has been simply to show how one sort of emergent feature can avoid various difficulties inherent in supervenience treatments. What we have thus shown is the possibility of property instances being emergent, free from the difficulties stemming from the exclusion and downward causation arguments. Even if there were no actual examples of fusion, the account of emergence given here would be useful because it provides a coherent account of a particular kind of emergence that is devoid of the mysteries associated with earlier attempts to explicate the concept. There is, of course, the further question of whether our world contains examples of emergent property instances. The answer to this is a reasonably confident "yes." I shall here only sketch the form that such examples take, referring interested readers to more detailed sources for a richer description.

It frequently has been noted that one of the distinctive features of quantum states is the inclusion of nonseparable states for compound systems, the feature that Schrödinger called "quantum entanglements."[21] That is, the composite system can be in a pure state when the component systems are not, and the state of one component cannot be completely specified without reference to the state of the other component. Furthermore, the state of the compound system determines the states of the constituents, but not vice versa.[22] This last fact is exactly the reverse of what supervenience requires, which is that the states of the constituents of the system determine the state of the compound, but when the supervening properties are multiply realizable, the converse does not hold. I believe that the interactions which give rise to these entangled states lend themselves to the fusion treatment described in the earlier part of this paper, because the essentially relational interactions between the "constituents" (which no longer can be separately individuated within the entangled pair) have exactly the features required for fusion. One might be hesitant to use quantum entanglements as an argument by themselves because of the notorious difficulties involved in providing a realist interpretation for the theory. But what seems to me to be a powerful argument in favor of the existence of these emergent features is that these quantum entanglements are the source of macroscopic phenomena that are directly observable. In particular, the phase transitions that give rise to superconductivity and superfluidity in helium are a direct result of nonseparable states.[23]

It is a question of considerable interest whether, and to what extent, fusion occurs in other areas. It would be a mistake to speculate on such matters, because the existence of such interactions is a contingent matter, to be settled by scientific investigation. There is indeed, within that part of metaphysics

---

[21] The discussion of nonseparability goes back at least to Schrödinger 1935. D'Espagnat 1965 is another early source and more recent discussions can be found in Teller 1986, Shimony 1987, French 1989, Healey 1991, and d'Espagnat 1995, among many others.

[22] See, e.g., Beltrametti and Cassinelli 1981, 65–72.

[23] See Shimony 1993, 221.

that can be (partially) naturalized, an important but neglected principle: *certain metaphysical questions cannot be answered (yet) because we do not know enough.* On the basis of this principle, those readers who want an answer as to whether, for example, mental phenomena are emergent in the above sense will, I am afraid, have to be patient.

There is one other way in which fusion can occur, and it is neither a matter of speculation nor something directly amenable to empirical inquiry. To see it, one simply needs to be reminded of something that has been lost in the avalanche of logical reconstructions of causation and other concepts in this century. It is that singular causal interactions between property instances, construed realistically, provide "horizontal" examples of the kind of novelty that has here and elsewhere been discussed in "vertical" terms. By "construed realistically," I mean "taken to be *sui generis* features of the world, the properties of which are fundamentally misrepresented by reductive analyses or (Humean) supervenience treatments of causation."[24] This is not the place to persuade readers of the benefits of the realist singular view,[25] so I shall simply note that the issues of "horizontal" and "vertical" novelty are connected, for the explaining away of the latter, especially by supervenience, tends to gain its plausibility from a sparse ontology of space-time points possessing a restricted set of primitive physical properties. If you believe, in contrast, that solid-state physics (for example) is more than just advanced elementary particle physics, you will begin to ask how phenomena from the two fields interact. You should then be prepared to find that emergence may be complicated, but that it is neither mysterious nor uncommon.[26]

Acknowledgments

Previous versions of this paper were read at Virginia Polytechnic Institute, Duke University, the University of Virginia, the University of Pittsburgh, the British Society for the Philosophy of Science, and an IUHPS meeting in Warsaw. Comments and suggestions from those audiences and the two anonymous referees from *Philosophy of Science* were very helpful in improving the paper. I am also grateful for conversations and correspondence with Robert Almeder, James Bogen, Richard Burian, John Forge, David Henderson, Martin

---

[24] I focus on causal interactions here only because of their familiarity. Other kinds of interactions can, one assumes, produce genuine novelty.

[25] For a partial account, see Humphreys 1989, sec. 25.

[26] Since this paper was first drafted in 1991 I have realized that the term "fusion" has a standard use in the mereological literature that is almost opposite to its use here. I believe that my use is better justified etymologically, but mereology was there first. The reader is hereby advised never to confuse the two uses.

Jones, Jaegwon Kim, James Klagge, Ken Olson, Fritz Rohrlich, and Abner Shimony. Research for this paper was conducted partly under NSF grant SBR-9311982 and the support is gratefully acknowledged.

References

Beltrametti, Enrico G., and Gianni Cassinelli. 1981. *The Logic of Quantum Mechanics.* Reading, MA: Addison-Wesley.

d'Espagnat, Bernard. 1965. *Conceptions de la physique contemporaine.* Paris: Hermann.

———. 1995. *Veiled Reality.* Reading, MA: Addison-Wesley.

French, S. 1989. "Individuation, Supervenience, and Bell's Theorem." *Philosophical Studies* 55: 1–22.

Healey, Richard A. 1991. "Holism and Nonseparability." *Journal of Philosophy* 88: 393–421.

Humphreys, Paul. 1989. *The Chances of Explanation.* Princeton, NJ: Princeton University Press.

Kim, Jaegwon. 1984. "Concepts of Supervenience." *Philosophy and Phenomenological Research* 45: 153–76.

———. 1988. "Supervenience for Multiple Domains." *Philosophical Topics* 16: 129–50.

———. 1992. "'Downward Causation' in Emergentism and Nonreductive Physicalism." In *Emergence or Reduction? Essays on the Prospects of Nonreductive Physicalism*, edited by Ansgar Beckermann, Hans Flohr, and Jaegwon Kim, 119–38. New York: Walter de Gruyter.

———. 1993. "The Nonreductivists' Troubles with Mental Causation." In *Mental Causation*, edited by John Heil and Alfred Mele, 189–210. Oxford: Oxford University Press.

Lewis, David. 1986. "Causation" and "Postscripts to 'Causation.'" In *Philosophical Papers*, vol. 2, 159–213. Oxford: Oxford University Press.

Schrödinger, Erwin. 1935. "Discussion of Probability Relations between Separated Systems." *Proceedings of the Cambridge Philosophical Society* 31: 555–63.

Shimony, Abner. 1987. "The Methodology of Synthesis: Parts and Wholes in Low-Energy Physics." In *Kelvin's Baltimore Lectures and Modern Theoretical Physics*, edited by Robert Kargon and Peter Achinstein, 399–423. Cambridge, MA: MIT Press.

———. 1993. "Some Proposals Concerning Parts and Wholes." In *Search for a Naturalistic World View*, vol. 2, 218–27. Cambridge: Cambridge University Press.

Teller, Paul. 1986. "Relational Holism and Quantum Mechanics." *British Journal for the Philosophy of Science* 37: 71–81.

Yablo, Stephen. 1992. "Mental Causation." *Philosophical Review* 101: 245–80.

{ 6 }

Emergence, Not Supervenience

6.1. Introduction

Supervenience has for years been the tool of choice for antireductionists. For reasons that are now familiar—the inability to derive theories about a higher-level domain from theories about a more fundamental domain, either because of syntactic complexities or the lack of the appropriate concepts in the lower domain; the multiple instantiability of concepts in the mental realm, such as "pain," and the resulting implausibility of reducing these concepts to unitary concepts in a lower realm; the mismatch of natural kind terms within laws at different levels, and so on—reduction has been out and supervenience has been in. This position is only half right. Reduction is still not an option, but supervenience is no good either. It is a notion that is empty of any scientific content, and what antireductionists need in its place is emergence. The latter idea can properly capture the picture of distinctively different layers of the world in which antireductionists believe. As always, there is a price to pay for adopting an emergentist view. In certain cases one must give up a number of metaphysical views that, for good or ill, have proven attractive to many. One of these is ontological minimalism.

Ontological minimalism runs roughly like this: (a) there is a relatively small set of fundamental constituents of the world, (b) to individuate these we need only intrinsic (i.e., nonrelational) properties, and (c) all the nonfundamental individuals and properties are composed of or from these fundamental entities. Ontological minimalism is an attractive view. Indeed, if one accepts a well-known set of philosophical doctrines, it comes out almost right. Chief among these doctrines are (1) giving primacy to logical reconstructions of ordinary and scientific concepts, and (2) a broadly Humean account of causal relations. If, in contrast, one thinks that these two doctrines (and related ideas) are wrong, then ontological minimalism will seem to be a perverse misrepresentation of the way our world happens to be. This is not the place to argue in detail against those two doctrines (arguments for their rejection can be found in my

1995 and 1997a).[1] Suppose that one did accept the rough picture provided by these two doctrines (which I do not). Then the core belief that underlies onto-logical minimalism is that nonfundamental entities are *nothing but* collections of fundamental entities. What is important for present purposes is that onto-logical minimalists commonly interpret the "nothing but" as either "aggregate" or "supervenient upon."[2]

6.2. Supervenience

Here I shall try to show that supervenience is similarly inadequate, albeit in a different way.

Despite the fact that many supervenience accounts are nonreductive in in-tent, there is a strong residue of suspiciously reductionist terminology within many of them. There is the "If A supervenes upon B, then A is nothing but B" talk of course. In addition, there is the idea that if A supervenes upon B, then because A's existence is necessitated by B's existence, all that we need in terms of ontology is B.[3] This is a core belief that motivates physicalism—all one needs is the ontology of physical objects and properties, then everything else supervenes upon those. My criticisms will apply, mutatis mutandis, to weak and global supervenience.

One standard definition of strong supervenience is this:

Definition: A family of properties M strongly supervenes on a family N of properties iff, necessarily, for each x and each property F in M, if F(x) then there is a property G in N such that G(x) and necessarily if any y has G it has F. (Kim 1993, 65)

Such definitions have little content until the kind of necessitation involved has been given. So let us consider nomological necessity and logical (or

---

[1] I am here accepting only for expository purposes the idea that there exist distinct levels of individuals and properties. Since most advocates of supervenience believe in this sort of layered on-tology, it would be a distraction to deny it here. So, for example, I here go along with the view that atoms occupy a lower level than do simple inorganic molecules, which in turn occupy a lower level than do complex organic molecules. These strata then provide a separation between properties of atoms, simple inorganic molecules, and complex organic molecules. Arguments against the levels view can be found in Humphreys 1995 and 1997a.

[2] Interpretations of "A supervenes on B" as "A is nothing but B" can be found (in so many words) in Armstrong (1989, 56): "if it supervenes, I suggest, it is not distinct from what it supervenes upon," and in Lewis (1986, 14): "all there is to the picture is dots and non-dots at each point of the matrix. The global properties are nothing but patterns in the dots."; "Symmetry is nothing but a pattern in the ar-rangement of dots" (Lewis 1986, 15). See also Rosenberg 1997 for similar sentiments about reduction.

[3] Again, Armstrong (1989, 100): "The relation supervenes upon the terms. . . . That, I think, makes the relation an ontological free lunch."

metaphysical) necessity (for both operators) as natural interpretations of the modal operators.[4]

*Super*venience, as its name implies, is standardly taken to be a relation between two collections of properties, one at a higher level than the other. Yet there is nothing in the definition itself that requires the relation to hold only between different levels, and this raises an interesting question: why is supervenience taken to be revealing as an interlevel relation but has no significant applications as an intralevel relation? Or, to put it another way, in what way, if at all, does the existence of a supervenience relation between two sets of properties explain why the members of the supervening set have an inferior ontological status than do the members of the subvening set? (And have no doubt that they carry an inferior status, for how else could physicalism have the appeal that it does if physical states were not the only legitimate ontology?) To see why this is a problem, let us, as it were, turn the supervenience relation on its side and apply it within a given level. First, consider the causal relation. Let M be a set of "effect properties," and N be a set of "causal properties," i.e., M consists in properties, instances of which are characteristically found as effects of instances of properties in N. For a simple example, take M as expansion in volume, and N as properties that cause expansion, such as heat, increase in pressure, and so on. Then, with the first interpretation of the necessities in the definition as nomological necessities, it is easily seen that in every possible world with the same laws as ours, if something expands, then there is some other property such as heat that applies to that thing and this nomologically necessitates its expansion.

Now, no one will agree that expansion is "nothing but" application of heat. Why not? One plausible explanation is that both heat and expansion are already considered to be respectable physical properties, and so there is no need to bring in supervenience to (putting it slightly pejoratively) "explain away" the supervenient properties. A second, but quite different, interpretation of supervenience, to the effect that the supervening properties are real, but have some secondary sort of status that is necessarily dependent upon their subvenient properties,[5] does not seem very promising either, because both heat and expansion are ordinarily taken to be equally real.

In the expansion case, it is indisputable that an instance of expansion is nomologically dependent upon (an instance of) application of heat but this is not the issue under discussion here; it is whether the properties themselves are so dependent. Of course, one already knows what the difference is between

---

[4] In the discussion after the symposium, Andrew Melnyk suggested that some form of necessitation other than nomological or metaphysical might well be appropriate for supervenience relations. There was not enough time before submitting this paper to explore this option.

[5] This is the interpretation that seems to be favored by Kim throughout much of his 1993, although he now seems to have doubts about the tenability of various supervenience views.

the interlevel and the intralevel applications of supervenience. In the case of interlevel applications, the inferior status of the supervening properties is nothing more than their initially having been considered to be less respectable than their subvenient bases, which is why the supervenience relation was constructed in the first place. Yet this clearly reveals that whatever the supervenience relation does, it does not, in itself, answer our question. That is, *there is nothing in the nature of the supervenience relation itself that will explain why "vertical" uses are appropriate and "horizontal" uses are not.*

Take a second case, one involving logical or metaphysical necessitation. Here let M be the set consisting only of the property "is triangular" and N the set consisting only of the properties "closed," "three-sided," "Euclidean," and "polygon." Then it is again obvious that M supervenes upon N. To say that M is nothing but N will not concern you if you subscribe to the position that metaphysically equivalent properties are identical. Although some philosophers have accepted this, many have found it to be quite implausible, and those of us who reject it in doing so also reject the idea that the property of being triangular has some sort of secondary, dependent, status.

In both of the examples I have described, we have good reason to hold that the supervenient property has some independent standing, irrespective of the fact that there are necessitation relations holding between it and other properties. With interlevel applications of supervenience, I doubt that anyone holds that the existence of a supervenience relation is itself a reason for ascribing some sort of secondary status to the supervenient property—this is already assumed. Mental properties, aesthetic properties, moral properties are antecedently taken to be less desirable than physical properties. So the question that remains is this: what is it that makes the supervenience relation admissible as an interlevel relation, but renders it wrongheaded as an intralevel relation? It cannot be simply a prejudice in favor of physical properties, for that would not explain why it gives unacceptable answers for geometrical properties. I believe that the answer is this: supervenience is acceptable as a consistency condition on the attribution of concepts, in that if A supervenes upon B, you cannot attribute B to an individual and withhold A from it. If aesthetic merit supervenes upon just spatial arrangements of color on a surface, and you attribute beauty to the Mona Lisa, you cannot withhold that aesthetic judgement from a perfect forgery of the Leonardo painting. But supervenience does not provide any understanding of *ontological* relationships holding between levels. For that emergence is required.

6.3. Possible Criteria for Emergence

Perhaps the most characteristic feature of emergent properties is that they are novel. Novelty can mean many things, but the simplest and crudest interpretation is that a previously uninstantiated property comes to have an instance.[6] Clearly, it is important how this novel instance came about. If the new value of, say, mass comes about by mere rearrangement of existing matter (by aggregating bits of smaller mass, or slicing up a larger mass) then we say that the "novel" property instance was there all along, it just was not instanced by the object that now has it. (In the aggregate case, it was a collection of spatially separated objects that had the "new" mass; in the slicing case, then it was a part of the object that had it). The novelty criterion for emergence would then be violated, and so this turns out to be no case of emergence at all. Suppose in contrast that the new mass value occurred spontaneously (from nothing), or was the result of (spontaneous) conversion of energy to mass. Here there is more of a sense of novelty, emergence from something else or from nothing. But this is a boring kind of emergence, and not the kind of case in which we are interested, for what we want from emergence is not a new value of an existing property, but a novel kind of property. This then gives us the second characteristic feature of emergent properties, which is that they are *qualitatively different* from the properties from which they emerge.

Four other features of emergent properties suggest themselves. The third is that an emergent property is one that could not be possessed at a lower level— it is logically or nomologically impossible for this to occur. The fourth is that different laws apply to emergent features than to the features from which they emerge. This antireductionist view has been suggested by the physicist P. W. Anderson: "Reductionism is the thesis that all of the animate and inanimate matter of which we have any detailed knowledge is assumed to be controlled by the same set of fundamental laws" (Anderson 1972, 393). We should then have that entities of type B are emergent from entities of type A iff entities of type B have type A entities as constituents and there is at least one law that applies to type B entities that does not apply to type A entities.

This is one sense that gives us the novelty—before the property instances interacted there never had been an instance of the new property. We might want to add that the property instance occurs at a higher level, or that the property instance is attached to a new kind of entity. That is, a hydrogen atom is an emergent entity—it is not just an electron and a proton put in a new spatial arrangement, but there is an essential interaction between them.

And the sixth feature is that emergent properties are holistic in the sense of being properties of the entire system rather than local properties of its constituents.

A nonrelational property of a whole is emergent if it supervenes upon but does not reduce to the non-relational properties of its parts (Teller 1992, 142).

---

[6] This sort of case is canvassed, and rejected in Teller 1992.

6.4. Some Examples

I do not suggest that any emergent phenomenon must satisfy all of these criteria, for there is a wide variety of ways in which emergence can occur. In my 1997b, I provide an account of one kind of emergence in terms of so-called "fusion" operations, and it turns out that it fits the most widely canvassed examples of emergence, those involving quantum entanglements. This sort of emergence directly satisfies our fifth and sixth criteria above, and when it is the basis for macroscopic phenomena such as superconductivity and superfluidity, it will satisfy at least criteria one, two, and four as well.[7] In fact, we can look to macroscopic phenomena for examples of other kinds of emergence too.

The idea that emergence can involve a qualitative difference in properties even though the object possessing those properties is composed from more elementary objects occurs in what is known as the theory of macroscopic systems. A macroscopic system can be defined as one whose equation of state is independent of its size (Sewell 1986, 4). An ideal macroscopic system is one containing an infinite number of particles, the density of which is finite. A real macroscopic system is one that is sufficiently large that its equation of state is empirically indistinguishable from an ideal macroscopic system (see Sewell 1986, 4). These definitions are purely objective and do not rely on a characterization of "macroscopic" by way of reference to human observers. One of the principal features of macroscopic systems in that sense is that they display qualitatively different properties than do the microcomponents composing them and that they are holistic. They thus satisfy our second and sixth criteria above. To illustrate this, here is a typical quotation:

> Macroscopic systems enjoy properties that are qualitatively different from those of atoms and molecules, despite the fact that they are composed of the same basic constituents, namely nuclei and electrons. For example, they exhibit phenomena such as phase transitions, dissipative processes, and even biological growth, that do not occur in the atomic world. Evidently, such phenomena must be, in some sense, *collective*, in that they involve the co-operation of enormous numbers of particles: for otherwise the properties of macroscopic systems would essentially reduce to those of independent atoms and molecules. (Sewell 1986, 3)

Also:

> The quantum theory of macroscopic systems is designed to provide a model relating the bulk properties of matter to the microscopic ones of its

---

[7] Whether it satisfies the third criterion depends upon how the levels are determined, something I have put aside here.

constituent particles. Since such a model must possess the structure needed to accommodate a description of collective phenomena, characteristic of macroscopic systems only, it is evident that it must contain concepts that are *qualitatively* different from those of atomic physics. (Sewell, 1986, 3)

This kind of talk is reminiscent of Rohrlich's conceptual irreducibility (Rohrlich 1997),[8] the main point being that the emergent properties *cannot* be possessed by individuals at the lower level because they occur only with infinite collections of constituents. Some of the most important cases of macroscopic phenomena are phase transitions, such as the transition from liquid to solid. This transition is not exhibited by the microcomponents of the liquid (or the solid) since the individual components are the same in each phase. It is their collective relationship to each other that changes across the (usually discontinuous) phase transition. Thus, it is the interactions between the constituents that makes for the qualitatively different macroscopic behavior. An example of this kind of interactive behavior is the spontaneous ferromagnetism that occurs below the Curie temperature, which is absent above that phase transition point and which occurs because of spin interaction coupling between adjacent particles on the lattice. It might appear that such cases are no different from more familiar ones, wherein the properties of water molecules, say, are different from the properties of the hydrogen and oxygen atoms that compose them. Essentially macroscopic phenomena are different from these, however, as the following example shows.

There are many qualitatively distinct macroscopic phenomena associated with ferromagnets and similar systems, but one will suffice here. If one takes a ferromagnet whose Hamiltonian is spherically symmetric, then below the Curie temperature the system is magnetized in a particular direction, even though because of the spherically symmetric Hamiltonian, its energy is independent of that specific direction. This divergence between the symmetry exhibited by the overall system and the symmetry exhibited by the laws governing its evolution is an example of spontaneous symmetry breaking. We have here a case where there is a distinctively different law covering the $N \to \infty$ system than covers its individual constituents. This is exactly the kind of difference of laws across levels of analysis that we noted earlier as one criterion of a genuinely emergent phenomenon.

It is important to emphasize the qualitatively different nature of these macroscopic phenomena, because it is not simply a result of our inability to provide an explicit Schrödinger equation for the macroscopic system (let alone solve it). It is a consequence of considering the macroscopic system as a system composed of an infinite number of particles, with boundary conditions that are macroscopic

---

[8] Although I find Rohrlich's conception of emergence too psychological for my taste.

in form. This means that the system is *necessarily* nonmicroscopic, in that it is inconsistent to consider a collection of infinitely many particles in conceptually the same way as an atomistic component, which is essentially singular.

These sorts of examples are persuasive, but they do of course rely on idealized models of real systems. That is in the nature of the examples and should not count against them as examples showing that emergent phenomena are not inescapably mysterious, are perhaps common, occur within indisputably *physical* systems rather than just biological or psychological systems, and can be given a treatment that is far more detailed than any supervenience account can even approach.

6.5. Conclusion

I believe that the arguments given here reveal three things. First, that emergent properties are probably quite common in the physical realm. If this is true, it will be likely that there will be no sharp boundary between the physical level and other levels. In turn, this will require refinement in terms of what we mean by physicalism, and in our reasons for holding it. Second, that the existence of these detailed models at the physical level dissolves the air of mystery that has traditionally surrounded emergentism and has led clearheaded philosophers to stay away. Finally, that the level of detail available in these models makes the use of supervenience relations seem simplistic. And it is for these three reasons at least that we need emergence, not supervenience.

References

Anderson, Philip W. 1972. "More Is Different." *Science* 177: 393–96.

Armstrong, D. M. 1989. *Universals: An Opinionated Introduction.* Boulder, CO: Westview Press.

Humphreys, Paul W. 1995. "Understanding in the Not-So-Special Sciences." *Southern Journal of Philosophy* 34, Supplement: *Spindel Conference 1995*, 99–114.

_____ 1997a. "Aspects of Emergence." *Philosophical Topics* 24 (1): 53–70.

———. 1997b. "How Properties Emerge." *Philosophy of Science* 64: 53–70.

Kim, Jaegwon. 1993. *Supervenience and Mind.* Cambridge: Cambridge University Press.

Lewis, David K. 1986. *On the Plurality of Worlds.* Oxford: Blackwell.

Rohrlich, Fritz. 1997. "Cognitive Emergence." *Philosophy of Science* 64: S346–S358.

Rosenberg, Alex. 1997. "Can Physicalist Antireductionism Compute the Embryo?" *Philosophy of Science* 64: S359–S371.

Sewell, G. L. 1986. *Quantum Theory of Collective Phenomena.* Oxford: Clarendon Press.

Teller, Paul. 1992. "A Contemporary Look at Emergence." In *Emergence or Reduction? Essays on the Prospects of Nonreductive Physicalism,* edited by Ansgar Beckermann, Hans Flohr, and Jaegwon Kim, 139–53. Berlin: Walter de Gruyter.

Synchronic and Diachronic Emergence

7.1. Introduction

Approaches to emergence are often divided into two broad categories, those of diachronic and synchronic emergence. The first approach primarily, but not exclusively, emphasizes the emergence of novel phenomena across time; the second emphasizes the coexistence of novel "higher level" objects or properties with objects or properties existing at some "lower level." The purpose of this article is to explore some relations between the two kinds of emergence. In particular, I shall argue for the following theses:

(1) Although I remain optimistic that we shall eventually find a unifying framework that explains why synchronic and diachronic emergence both count as emergence in some more general sense,[1] the two kinds of emergence at present remain conceptually distinct. In particular, the current criteria for synchronic emergence are not sufficient for a state or property instance to count as emergent because the historical development of a system's dynamics is often crucial to the system's terminal states being emergent. There can be two instances of the same state, one of which is emergent and the other not, the difference being solely in the way in which they were generated. This feature introduces an ineliminable element of historicity into considerations of emergence.

(2) The account that goes under the name "weak emergence" (Bedau 1997, 2003) captures much of what is important about computational forms of diachronic emergence. However, if the criteria for weak

---

[1] For some criteria shared by different accounts of emergence, see Humphreys 2006.

emergence are supplemented, they can more accurately capture the class of computationally emergent phenomena. I supply the elements of a friendly amendment that further constrains the conditions of application for weak emergence.

(3) Some serious account of conceptual emergence will be required to fully account for what I call "pattern emergence." I do not have much that is useful to say about conceptual emergence here but I shall briefly discuss how the relevant issues fit within existing philosophical debates.

7.2. Pattern Emergence

A common type of emergence involves the appearance in a system of novel structure that results from the temporal evolution of the system. For brevity, I shall call this phenomenon "pattern emergence." Pattern emergence is a common phenomenon in computational models such as agent-based simulations and cellular automata, and it is widely agreed within the complexity theory literature that these patterns count as examples of emergent phenomena. Although these claims should not be taken uncritically, they can serve as a plausible starting place for philosophical analysis.

Some preliminary remarks: because our first two theses involve counterexamples to general claims about emergence, we can establish each of them by restricting ourselves to examples drawn from the area of pattern emergence. In contrast, the positive remarks I make are restricted to the particular types of cellular automata models considered. In particular, the patterns I discuss here all have spatial structure. The task of extending our understanding of structured outputs from spatial patterns to different kinds of regularities is difficult and requires an extended treatment of conceptual innovation that I do not yet have. Next, although pattern emergence can be considered either from the point of view of the models that generate the patterns or from the perspective of the real systems being modeled I restrict myself to consideration of the models because their structure is ordinarily better understood than are the real-world phenomena. Indeed, one advantage of these examples is that there is no room for speculation about the details of the examples because the algorithms that underlie them are explicitly given, thus avoiding the speculative mist that surrounds many discussions of emergence in the philosophy of mind. Finally, pattern emergence has some additional philosophical interest because it illuminates the issue of what used to be called methodological individualism in the social sciences. Pattern emergence occurs because of a "bottom up" set of processes starting with interactions between the individual constituents of a larger entity and not with a centralized "top down" set of rules or laws that govern the behavior of individuals. I shall not, however, explore that aspect in this paper.

7.3. Cellular Automata

The concept of weak emergence has been developed in two papers by Mark Bedau (1997, 2003). It is most easily understood with the aid of some examples. Consider a two-dimensional cellular automaton (CA). The cells on an infinite discrete grid[2] are labeled by their coordinates (i, j) and each possesses a discrete N-valued property (for concreteness think of these states as colors) the value of which constitutes the state of the cell. The state of every cell is simultaneously updated at each time step by a deterministic rule that is a function of the current state of the cell and the states of its immediate neighbors.[3] Over a wide range of initial states of the CA, appropriate updating rules can produce randomly distributed arrays of colored cells, stable patterns that persist across time, and dynamic patterns that evolve over time.[4] We can take as canonical a very simple, albeit idealized example. The cells of the cellular automaton are either black or white and after a considerable number of computational steps using the rules that define the CA (what these are is irrelevant here) the pattern in figure 7.1 is displayed.

Call the geometrical property displayed here "bow-tie shaped." I shall examine weak emergence in detail in a moment, but the essence of the idea is that a state of a system is weakly emergent just in case that state can be produced only through a step-by-step simulation of the system. In other words, the process that leads up to the state is computationally incompressible. In yet other words, unlike the prediction of future solar eclipses for which the computational difficulty of prediction is almost independent of how far into the future the eclipse will take place, predictions of future states of computationally

---

[2] Finite grids can emulate infinite grids by using periodic (wraparound) boundary conditions.

[3] Using all eight immediate neighbors gives the Moore neighborhood. Von Neumann neighborhoods, in which the four diagonal neighbors do not influence the active cell, are not essentially different. I do not consider here sequential cellular automata in which the states of the cells are updated sequentially, rather than simultaneously.

[4] Cellular automata have the philosophically interesting feature that they serve as models for Humean worlds. The state of each cell is an intrinsic property of that cell, the cells and their states are, within the discrete space and time of CAs, point-like, and the rules are all local. Whether these Humean worlds are also worlds satisfying the conditions for Humean supervenience is not an issue I shall pursue here.

FIGURE 7.1 *Bow tie pattern in cellular automaton*

incompressible systems must run through each of the intermediate time steps between the initial state and the predicted state. Letting the computational model work out its own development is thus the only effective way to discover how the system's states evolve. The philosophical motivation for accepting this criterion as capturing a certain kind of emergence draws on the philosophical tradition that emphasizes the essential unpredictability of emergent phenomena. The work of C. D. Broad, for example, lies in the essential unpredictability tradition, although he, of course, did not make use of computational criteria.

7.4. Properties of Pattern Emergence

I shall now argue that the historical development of a pattern is essential to its status as an emergent entity. It is an essential feature of emergence that the emergent entity must emerge *from* something else. Consider a token of the bow-tie pattern that was generated by running the cellular automaton over n time steps. The process by means of which that token is generated is simply the iteration of the rules n times and it is from that process that the pattern emerges. Now suppose that I print an exact duplicate of that pattern using a bow-tie-shaped rubber stamp. It is another token of the same pattern, but that token is not emergent because it is generated instantaneously.

In fact, there is nothing from which the token emerges, there is just the pattern itself, which was produced at a time instant.[5] This reveals three things. Pattern emergence is an essentially historical phenomenon—whether an instance of a pattern is emergent or not depends essentially upon the process that generated it. It is therefore impossible to determine whether a pattern is emergent by looking only at synchronic relations between the pattern and the spatial array of elements that comprise the pattern. Compare this with what is claimed about synchronic relations such as supervenience or realization. In treatments of emergence that use supervenience relations, such as those of van Cleve (1990) and McLaughlin (1997), one is supposed to be able to determine by examining an instantaneous state of a system whether the higher-level property is emergent from the lower level. The reason for the impossibility of doing this in the present case and others like it lies in the fact that supervenience relations are about types, properties, universals, as they must be in order for them to capture more than the token-token identity relations that both reductionists and antireductionists agree hold between the levels related by these supervenience relations. Furthermore, it is in the very nature of the necessitation relation embedded in supervenience relations that identical bases give rise to identical supervenient features. As a result, no synchronic account of emergence based on supervenience relations of which I am aware is capable of differentiating the two instances of the pattern. In addition, the argument shows that *pattern emergence is about tokens or instances of patterns*, not about types. In many discussions of emergence, the position is that if a given property is emergent in any of its instances, it is emergent in all of its instances. This, as we have just seen, is false. Two cases of the same pattern type may be emergent in one instance and not emergent in another.

7.5. Weak Emergence

Weak emergence is defined in the following way: "Assume that P is a nominally emergent property possessed by some locally reducible system S. Then P is weakly emergent if and only if P is derivable from all of S's micro facts but only by simulation" (Bedau 2003). Some elaboration: A nominally emergent

---

[5] If the response is that the pattern emerges from the causal interaction between the rubber stamp, the ink, and the paper, this will lead to admitting most cases of causal interaction as producing emergent entities, a position which would make emergence a very common phenomenon and hence would count for most people against that position. See also footnote 8.

property is a property that can be possessed at the macro level but is in principle incapable of being possessed at the micro level. For example, individual cells in a CA can only be square and so the property "is bow-tie shaped" is nominally emergent. A locally reducible system is, roughly, a system in which all of the macroproperties are structural properties, i.e., the state of a micro entity consists of its location and intrinsic properties whereas the state of a macro entity is simply the aggregate of the states of its microconstituents together with their spatial relations. A system consisting of the set of output states of a cellular automaton is locally reducible. A simulation, in the special sense used here, is a step-by-step process that replicates the time development of the system at the micro level. The bow-tie example is thus a case of weak emergence, assuming that the processes leading to the generation of the pattern are computationally incompressible.

The definition of weak emergence succinctly captures a great deal of what is considered emergent within the fields of dynamical systems theory and complexity theory but it needs to be supplemented in two ways. As Cyrille Imbert (2007) has pointed out, Bedau's definition places no restriction on the kind of end-state property that can weakly emerge from a simulation and argues that this is a defect in the definition. Imbert focuses on a contrast between what he calls "deceptive properties" and "target properties"; roughly properties that are (nothing more than) conjunctions of microstates and what we think of as predicates picking out genuine macro-level properties. He takes our acceptance of the latter and rejection of the former as emergent to be intuitively evident, but it is worth spelling out explicitly why this is. Under most information-theoretic characterizations of randomness, such as Kolmogorov complexity, specification of random arrays can be given only by a massive, cell by cell, conjunctive specification of the microstates. A requirement that the end-state pattern be nonrandom can be justified in two ways. Suppose first that the system transitions from an initial random configuration to a final random configuration. The end state would then violate the novelty criterion for emergence. Although random patterns are token distinct, qua random they are type identical and so no new property has been produced.

If instead the system makes a transition from an initially structured state to a final random state we can appeal to the reasons that motivate interest in self-organizing systems to reject this as a case of emergence. The basic idea behind self-organization is that large-scale structure spontaneously emerges in a dynamic system of interacting constituents (i.e., the structure emerges without the need for external interventions) solely in virtue of local interactions between those constituents, and this structure is not accidental in the sense that were the system to be restored to its initial state or a qualitatively equivalent state and the microdynamics rerun, similar large-scale structure would, with a high degree of probability, reappear.[6] It is the apparent ability to run counter to the general trend produced by the Second Law of Thermodynamics that makes the transition from disorder to order so interesting. Order emerging from disorder strikes us as important because it is unusual; disorder arising from order as uninteresting because it is so common. So for the purposes of this paper, I am going to assume that some sort of structure is a necessary condition for a pattern to count as emergent and that a random array does not count as structured. There is a spectrum of possibilities between a fully random array and a fully structured array and so it is unlikely that pattern emergence will be a dichotomous phenomenon. Indeed, it seems artificial to insist that emergence is an all-or-nothing property of a system and in this respect it is similar to self-awareness, which is also plausibly a matter of degree.[7]

In addition, imposing the randomness condition prevents trivializing the definition of weak emergence, simply because the maintenance of a stable random pattern is too easy to achieve.[8] For example, two-dimensional random patterns can be generated on finite grids by running through the coordinates and randomly assigning a color to each cell of the grid, but if such things count as emergent then so would the end result of any random process, and that is too generous. We now arrive at thesis (2). Random, unstructured states count as weakly emergent if they can be generated only by simulation just as much as do highly structured patterns. Because of these arguments, I suggest that we add the following clause to the definition of weak emergence, making pattern emergence a restricted form of weak emergence:

P is a nonrandom property of the system S that is distinct from any property possessed by the initial state of S.

7.6. Microstable and Microdynamic Patterns

If synchronic features are not sufficient for pattern emergence, are they necessary? The simple answer is "No," but as we shall see, it is hardly satisfactory to leave it at that.

---

[6] Self-organization is almost always a statistical property of a system. In many models, unusual initial conditions will prevent the structures from emerging.

[7] One interpretation of Wimsatt 2007 is that it is a sustained argument for emergence not being a dichotomous property.

[8] A commonly voiced opinion is that in order to be acceptable, a definition of emergence should not make emergent phenomena too common. I have some sympathy for that view, but it requires more than an appeal to intuitions. The requirement that the pattern be nonrandom is sometimes motivated by the belief that emergent phenomena occur in the region intermediate between completely random behavior and completely structured behavior, a region including the much-publicized "edge of chaos." I shall not consider that motivation here.

Within the domain of pattern emergence, we can identify two broad types of patterns. The first occurs when a computational process leads to a nonrandom pattern and the constituents of the pattern remain fixed. Call this a *microstable pattern*. The second kind consists in the appearance of a nonrandom pattern and that pattern remains invariant under a specified class of dynamic substitutions of the pattern's constituents.[9] Call any such pattern a *microdynamic pattern*. There are three distinct subtypes of microdynamic patterns that can result from the end-state dynamics of a system. In the first kind, the macrostructure is stable under dynamic microprocesses involving the same constituents over time. We can call this *recirculating autonomy*. Widely cited examples of this type of stability are Bénard convection cells and Couette flow. Bénard convection cells occur when a viscous fluid is heated between two horizontal rectangular plates and the fluid eventually forms into cylindrical convective rolls. Couette flow can be produced in a fluid placed between two cylinders rotating with different velocities and vortex rolls form when the velocity gradient exceeds a critical value. The distinguishing feature of recirculating stability is the emergence and persistence of a structure possessed by a fixed collection of entities, that structure remaining constant as the dynamics of the constituent microentities play out across some temporal interval.

In the second kind of stable pattern, which we can call *transient autonomy*, the macrostructure emerges and then persists through substitutions of microconstituents of the same type as the original, a case that applies to standing waves in river flows as well as, in some cases, to the products of social insects such as ants and termites. The standing wave persists even though countless different water molecules move through the three-dimensional region that contains the wave. The distinguishing feature of transient autonomy is thus the persistence of a stable structure as the original microconstituents are replaced by new ones of the same kind, this replacement usually occurring as a result of the dynamics at the micro level. The third kind of autonomy, which we can call *equivalence class autonomy*, occurs when the macrostructure is stable under replacement of microconstituents from a specified class of types of entities. An example of this would occur if the standing wave persisted when water was replaced by alcohol. Each successive type is a generalization of the previous type and microstable patterns are a degenerate case of recirculating autonomy.[10]

---

[9] In addition to Bedau 2003, this autonomy of emergent phenomena has been emphasized in Wimsatt 1994 and Batterman 2002. For example, Wimsatt 1994, p. 250 suggests that 'the dynamical autonomy of upper-level causal variables and causal relations—their apparent independence of exactly what happens at the micro- level—serves as a criterion for a new ontological level.

[10] There is a fourth kind of stability that I shall not discuss here, in which the pattern is stable under external perturbations. This is a feature that is characteristic of, for example, Reynold's boid models of the flocking behavior of birds. For details of the boid algorithms see http://www.red3d.com/cwr/boids/.

We thus have two quite distinct roles that the dynamics of microprocesses play in pattern formation and persistence. The first role involves the process that leads to the initial formation of the structured pattern. This is all that is involved in the generation of a microstable pattern. The second role involves the persistence of the pattern as the microdynamics of the system continue. This feature (normally preceded by the first) is present in all three kinds of pattern autonomy. The first of these roles is important for considerations of pure diachronic emergence, whereas the second introduces a different aspect of diachronic emergence—the persistence of an emergent pattern across time—that seems to require drawing upon concepts from both diachronic and synchronic emergence.

To illustrate this, in the case of our bow-tie example we can suppose that there are three different behaviors that bow-tie patterns exhibit after they appear. In the first type of behavior, the bow tie rotates clockwise by $\pi / 2$ radians at each time step. This provides a very simple example of recirculating autonomy and its salient feature is that the same four cells are involved in maintaining the pattern through the microdynamics. In the second type of behavior the bow tie translates across the infinite grid without colliding with another object. The translation consists in different elements of the grid successively taking on the pattern and is thus a very simple example of transient autonomy. Equivalence class autonomy could be modeled by a slightly more complex CA in which different colors instantiate the bow-tie shape at different time steps.

I now need to add a second example to the bow-tie case. This example is perhaps somewhat more complicated than is necessary, but it does represent a wide class of cases. Consider a two-dimensional 256×256 fourteen-state cellular automaton with von Neumann neighborhoods. The update rules for a Greenberg-Hastings (G-H) model of neural excitation and relaxation are

(1) If the target cell is in state m and an adjacent cell is in state 0, then the target cell goes into state 0. (Excitation of a cell by a neighboring excited cell)

(2) If the target cell is in state m and an adjacent cell is in state m + 1, for $0 \leq m \leq 13$, then the target cell goes into state m + 1. (Relaxation of the cell)

Otherwise, the target cell stays in its current state.

From an initial random state displayed in plate 7.2, the system evolves after 500 iterations into the pattern pictured in plate 7.3.

The pattern does not emerge until about 200 time steps have been completed and the full pattern takes about 500 steps to complete. Which initial random state occurs is largely irrelevant to the appearance of the final pattern, which is stable for further iterations of the algorithm. This is a paradigmatic example of a diachronically emergent microstable pattern.

I now note that the simple bow-tie pattern discussed earlier can be given an explicit definition in terms of intrinsic properties of individual cells and spatial relations between them:

> For any cells y, z, let the relation Adj(y, z) hold just in case y and z are diagonally adjacent, i.e., $y = (i, j) \Rightarrow z = (i+1, j+1)$ or $z = (i-1, j+1)$ or $z = (i+1, j-1)$ or $z = (i-1, j-1)$.

Let $\overline{Adj}(x, y)$ be a function that takes a pair of diagonally adjacent cells to the complementary diagonally adjacent squares. That is, if Adj(x, y), then $\overline{Adj}(x,y) = \pi/2(x,y)$, where $\pi/2$ is a clockwise rotation of the pair of cells $\overline{(x, y)}$ by $\pi/2$ radians about their point of intersection. Let $S(y) = 1$ if y is colored; $S(y) = 0$ if not. Then define B(x, y) iff $\overline{Adj}(x,y)$ & $S(x) = S(y) = 1$ & $S(f_1(\overline{Adj}(x,y))) = S(f_2(\overline{Adj}(x,y))) = 0$, where f_1, f_2 are the projection functions picking out, respectively, the first and second elements of the ordered pair.[11]

The Greenberg-Hastings example and the bow-tie example thus differ in one important respect. In the latter example, I introduced the familiar predicate "bow tie shaped" and then provided an explicit definition in terms of a simple logical combination of basic properties. When the pattern is more complicated, as it is in the Greenberg-Hastings example, it may well require, for purely pragmatic reasons, the use of a predicate, such as "spiral maze," that could, in principle, be given an explicit definition, although in practice this would need to be complicated in order to pick out the exact pattern involved. Aside from this difference in degrees of definitional complexity, is there any reason to think that the spiral maze property is a different property from a particular spatial pattern of cells? To address this point, some comments about multiple realizability are in order.

7.7. Pattern Emergence and Supervenience

The appeal to diachronic, "horizontal" emergence in our examples of microstable patterns lies at the heart of weak emergence. The autonomy of "vertically" autonomous microdynamically stable patterns seems to be a different kind of emergence. Do we therefore need to represent such patterns in terms of vertical determination relations or something similar? First, recall

---

[11] I am grateful to Mark Bedau for pointing out that my original definition was too loose because it allowed a uniformly black field of cells to contain multiple examples of the relation B. Although one might argue that this is acceptable, it is better to use the more complicated definition given here.

that we are limiting our discussion to computational models. This means that an abstractive step has already been made from the implementation level and so the fact that cellular automata can be realized electronically, with computer displays, or with cubes of colored Jell-O is irrelevant here. Those realizations are relevant to (causal) functionalism but not to the geometrical properties with which we are concerned.

Indeed, one can construct an argument to the effect that the microdynamic stability of structure involved in recirculating and transient autonomy occurs when multiple *instantiability* is involved rather than multiple realizability. These cases occur when a specific type of cell is involved in the realization of the emergent pattern, but which individuals are involved is irrelevant. Consider our first cellular automaton example. When recirculating or transient dynamics are involved and the bow-tie pattern rotates or moves across the grid, the pattern is instantiated at different time steps by different pairs of cells. Even though different-sized or different-colored cells could be used, the properties involved in the definition remain the same and are given in the explicit definition of B. Because multiple realizability is not involved in either of these cases of pattern emergence, this shows that type-type identities are possible for recirculating and transient autonomy. But this argument relies on a crucial assumption about pattern identity which becomes clear when we consider equivalence class autonomy. Suppose that instead of black and white cells in the bow-tie case, the CA has fourteen different states in which a cell can be, represented by fourteen different colors as in the G-H example. Then we need to have answers to questions such as, "Is the bow-tie pattern picked out only by a green-red pair of cells or will any single color or pair of colors in a bow-tie shape count?" If the pattern is not color specific, then not only do we have multiple realizability, but the example is trivialized, because any random array of cells will instantiate vast numbers of bow ties. This is easily seen from the initial random state of the G-H algorithm above. And in fact unless we have an argument which establishes that a green-red bow-tie shape constitutes a different pattern than does a blue-yellow bow-tie shape, it will not be possible to have an objective criterion for whether we have *any* of our three types of dynamic autonomy. The bow-tie example is far too simple to represent most cases of pattern emergence and the situation regarding pattern identity is, of course, much worse in other examples such as the G-H case not the least because we have no portmanteau term—no neat "bow tie" terminology—to capture the pattern in these cases. An appeal to supervenience will be vacuous simply because all patterns supervene on geometrical arrangements of cells.

We are now in the midst of a tangle of familiar issues, none of which have easy solutions. In Dennett 1991, the issue of what makes a pattern real is addressed through the design stance; the field of pattern recognition is devoted to such matters; complexity theory has grappled with the problem; and from its inception statistics has developed objective criteria to replace the perceptual judgments we intuitively make about what counts as a regularity.[12] If I am correct about the role that this kind of pattern persistence plays in diachronic emergence, then whether the criteria for pattern identity turn out to be objective, subjective, pragmatic, or based on some other ground, that feature will automatically carry over to computational diachronic emergence itself.

Acknowledgments

Thanks to Anouk Barberousse, Mark Bedau, Jean-Paul Delahaye, Jacques Dubucs, Serge Galam, Philippe Huneman, Cyrille Imbert, and Sara Franceschelli for helpful discussions on this topic. Thanks also to audiences at the June 2004 IHPST conference on the dynamics of emergence, Yale University, the April 2005 Rutgers philosophy of physics conference, and the 2005 APA Central Division meetings for pushing me to present some of the points more clearly.

References

Batterman, Robert W. 2002. *The Devil in the Details*. New York: Oxford University Press.
Bedau, Mark. 1997. "Weak Emergence." In *Philosophical Perspectives: Mind, Causation, and World*, edited by James E. Tomberlin, vol. 11, 375–99. Malden, MA: Blackwell.
———. 2003. "Downward Causation and the Autonomy of Weak Emergence." *Principia Revista Internacional de Epistemologica* 6: 5–50.
Dennett, Daniel. 1991. "Real Patterns." *Journal of Philosophy* 88: 27–51.
Humphreys, Paul. 2006. "Emergence." In *The Encyclopedia of Philosophy*, 2nd ed., edited by Donald M. Borchert. London: Macmillan.
Imbert, Cyrille. 2007. "Why Diachronically Emergent Properties Must Also Be Salient." In *Worldviews, Science, and Us: Philosophy and Complexity*, edited by Carlos Gershenson, Diederik Aerts, and Bruce Edmonds, 99–116. Singapore: World Scientific.
Li, M., X. Chen, X. Li, B. Ma, and P. M. B. Vitanyi. 2004. "The Similarity Metric." *IEEE Transactions on Information Theory* 50: 3250–64.
McLaughlin, Brian. 1997. "Emergence and Supervenience." *Intellectica* 25: 25–43.
Rueger, Alexander. 2000a. "Physical Emergence, Diachronic and Synchronic." *Synthese* 124: 297–322.
———. 2000b. "Robust Supervenience and Emergence." *Philosophy of Science* 67: 466–89.
Strogatz, Steven H. 1994. *Nonlinear Dynamics and Chaos*. Reading, MA: Addison-Wesley.

---

[12] An especially interesting recent attempt in information theory is Li et al. 2004. The kinds of patterns that occur in state space and that are treated in dynamical systems theory occur at a much higher level of abstraction than the geometrical patterns discussed here. For an interesting treatment of how such patterns are related to diachronic emergence, see Rueger 2000a, 2000b; also Strogatz 1994.

van Cleve, James. 1990. "Mind-Dust or Magic? Panpsychism versus Emergentism." *Philosophical Perspectives* 4: 215–26.

Wimsatt, William C. 1994. "The Ontology of Complex Systems." In *Canadian Journal of Philosophy* 24, supplementary vol. 20, edited by Mohan Matthen and Robert Ware, 207–74. Calgary: University of Calgary Press.

———. 2007. "Emergence as Non-aggregativity and the Biases of Reductionisms." In *Re-engineering Philosophy for Limited Beings: Piecewise Approximations to Reality*, 274–312. Cambridge, MA: Harvard University Press.

Computational and Conceptual Emergence

8.1. Introduction

It is useful to consider two complementary taxonomies for emergence. The *relational taxonomy* is based on relations between the emergent entity and the entities with respect to which it is emergent. There are three divisions in this taxonomy; the *inferential* approach, the *conceptual* approach, and the *ontological* approach.

8.1.1. THE RELATIONAL TAXONOMY

The inferential approach holds that an entity, such as a state or a property instance, is emergent with respect to a domain D if and only if it is impossible, on the basis of a complete theory of D, to effectively predict that entity or to effectively compute a state corresponding to that feature. The terms "complete theory," "impossible," and "effectively" serve as parameters in this approach that, when specified, produce a particular inferential account. To avoid an inferential account from being too easily satisfied, it should be accompanied by arguments that the domain D has some special features, such as universality as when D is the physical domain or closure as when D is the domain of a well-established science. Domains specific to a computational model can also be used. The primary relation involved holds between the emergent feature and the predictive basis. When the complete theory involved is one of limit science, this may count as a reason to speak of the underivability as absolute, but logically, the attribution of emergence is still relational. A classic account of this type can be found in C. D. Broad's 1925 book *Mind and Its Place in Nature*,[1]

---

[1] "The only peculiarity of [a transordinal law] is that we must wait till we meet with an actual instance of an object of the higher order before we can discover such a law; and that we cannot possibly deduce it beforehand from any combination of laws which we have discovered by observing aggregates of a lower order" (Broad 1925, 79).

where the impossibility condition is taken to be the impossibility of a formal derivation. Broad's account was formulated before the development of completeness theorems and formal theories of computability and it is, forgivably, unspecific in certain respects. More precise accounts of the relevant derivational relations underlie certain computationally based treatments of emergence, one of which, weak emergence, is discussed later. The work of Cosma Shalizi on identifying efficient predictors also fits into this tradition.

The conceptual approach to emergence often results from a situation in which some form of inferential emergence occurs. It maintains that an entity, such as a state or a property, is conceptually emergent with respect to theoretical framework F if and only if a conceptual or descriptive apparatus that is not in F must be developed in order to effectively represent that entity. This apparatus may include new predicates, a new law, or an entirely new theory, and the introduction of the new concepts will sometimes involve moving to a new scientific domain. The phrase "effectively represent" serves as a parameter for these approaches and may involve efficient prediction or description. For instance, the need to introduce the term "liquid" in order to effectively describe the behavior of a large collection of water molecules that has undergone a transition from the solid phase to the liquid phase is an example of conceptual emergence.[2] The primary relational aspect involved in conceptual emergence is one of conceptual irreducibility between the new and the old theoretical frameworks. Both the inferential approach and the conceptual approach apply to representations of the phenomena rather than directly to the phenomena themselves.

Turning to the last part of the relational taxonomy, the ontological approach considers emergent entities to be genuinely novel features of the world itself, where an entity is ontologically emergent with respect to domain D if and only if that entity is ontologically irreducible to entities in domain D. Candidates for ontologically emergent phenomena include entangled states in quantum mechanical systems and, in some approaches, properties that supervene on, but are not reducible to, properties in D. The former examples have the distinctive feature that the states of the individual systems before interaction do not determine the joint state after the interaction, whereas the joint state does determine the individual states. This feature makes compositional accounts of the joint system implausible. The primary relata in the ontological approach are the emergent feature and the features from which it emerges, but

---

[2] One concise statement of conceptual emergence is: "This, then, is the fundamental philosophical insight of twentieth century science: everything we observe emerges from a more primitive substrate, in the precise meaning of the term 'emergent', which is to say obedient to the laws of the more primitive level, but not conceptually consequent from that level" (Anderson 1995, 2020, quoted in Castellani 2002, 265 n. 6).

the relation can be one of functional irreducibility (Kim 1999), supervenience (McLaughlin 1997), fusion (Humphreys 1997a), or some other relation.

The three approaches are not mutually exclusive, although between them they do cover every contemporary account of emergence of which I am aware, except for mystical views that consider emergent phenomena to be immune from understanding. The fact that elements falling into each of these categories are relational is logically important, for rather than the criteria for emergence being applicable to an entity E in isolation, such as consciousness, one must consider a domain D or a framework F with respect to which E is emergent, in the sense that it is the relevant relations between E and D or F that determine the emergent status of E. The fact that some of these relations are negative relations, such as irreducibility, does not affect this general point.

I have not included in this taxonomy the approach that considers emergent features to be unexplainable because a striking characteristic of most contemporary scientific candidates for emergent phenomena is that we do have credible explanations of why those candidates have their emergent features. They are thus essentially different from the examples of an earlier era such as life, and they are also different from some contemporary candidates such as consciousness. No "natural piety" is need to accept modern, scientifically based examples of emergence.

8.1.2. THE TEMPORAL TAXONOMY

The second taxonomy, the *temporal taxonomy*, appeals to the differences between diachronic and synchronic emergence and is easier to articulate. Diachronic emergence primarily, but not exclusively, involves the emergence of novel phenomena from preceding phenomena as a result of a temporally extended process; synchronic emergence involves the simultaneous coexistence of novel "higher level" objects or properties with objects or properties existing at some "lower level." This division cuts across the tripartite division of the relational taxonomy; each one of those three divisions has diachronic and synchronic variants. For example, synchronic conceptual emergence can occur in theoretical contexts within which, according to a familiar tradition, Nagel's inhomogeneous reducibility (Nagel 1961, chap. 11) cannot be carried out whereas, when a material makes the phase transition to ferromagnetism, the brief but temporally extended relaxation into the global state given by **M** involves a diachronic process, the end state of which requires the introduction of a new concept, the long-range order, to fully describe. Supervenience accounts of emergence (e.g., van Cleve 1990; McLaughlin 1997) are usually synchronic; a diachronic account of emergence is given in Rueger 2000. Examples of each dual category are given in table 8.1.

TABLE 8.1 A sixfold taxonomy for emergence

	Essential Unpredictability	Conceptual	Ontological
Synchronic	Broad 1925	Anderson 1972	van Cleve 1990
Diachronic	Bedau 1997	Rueger 2000	Humphreys 1997a

8.2. Weak Emergence

A contemporary example of a computationally based inferential approach is the concept of weak emergence that has been developed by Mark Bedau (1997, 2003). "Assume that P is a nominally emergent property possessed by some locally reducible system S. Then P is weakly emergent if and only if P is derivable from all of S's micro facts but only by simulation" (Bedau 2003, 15).[3] Weak emergence is a diachronic, inferential approach to emergence and it is most easily understood with the aid of some examples. Consider a 2-dimensional cellular automaton. The cells on an infinite discrete grid[4] are labeled by their coordinates (i, j) and each possesses a discrete N-valued property (for concreteness, think of these states as colors) the value of which constitutes the state of the cell. The state of every cell is simultaneously updated at each time step by a deterministic rule that is a function of the current state of the cell and the states of its immediate neighbors.[5] Over a wide range of initial states of the cellular automaton, appropriate updating rules can produce randomly distributed arrays of differently colored cells, stable patterns that persist across time, and dynamic patterns that evolve over time.

Weak emergence fits into the inferential tradition because if a state of a system is weakly emergent then the existence of that state can be discovered for the first time only through a step-by-step simulation of the system, so that the system must go through all of the intermediate developmental states before the final state is displayed. In other words, the process that leads up to the state is computationally incompressible. In yet other words, unlike the prediction of future solar eclipses for which the computational difficulty of prediction is almost independent of how far into the future the eclipse will take place, predictions of future states of computationally incompressible systems must run through each

---

[3] An earlier account of emergence based on the unavoidability of simulation is Darley 1994.

[4] Finite grids can emulate infinite grids by using periodic (wraparound) boundary conditions.

[5] Using all eight immediate neighbors gives the Moore neighborhood. Von Neumann neighborhoods, in which the four diagonal neighbors do not influence the active cell, are not essentially different. The difference between sequential cellular automata in which the states of the cells are updated asynchronically and regular cellular automata in which cell states are updated simultaneously becomes relevant in section 8.3.3.

of the intermediate time steps between the initial state and the predicted state. Letting the computational model work out its own development is thus the only effective way to discover how the system's states evolve.

Adapting the approach in Rasmussen and Barrett 1995, the computational incompressibility aspect of weak emergence can be represented in the following way. Consider the kinematics of a system S that is modeled as a temporally discrete process. An updating function U takes the current state of the process and transforms it into a future state. There is then an essential difference between these two cases:

$$S(t+T) = U[S(t)],\tag{1}$$

for arbitrary T, and

$$S(t+T) = U_{\mathrm{T}}\Big[U_{\mathrm{T-1}}\big[\ldots\big[U_1[S(t)]\big]\ldots\big]\Big],\tag{2}$$

where U_i is the updating operator at time step i. In the usual case where U_i is fixed, we have $S(t+T) = U^T[S(t)]$.

8.3. Weak Emergence, Conceptual Novelty, and Some Reasons Why Weak Emergence Is Different from Conceptual Emergence

The definition of weak emergence succinctly captures a great deal of what is considered emergent within the fields of dynamical systems theory and complexity theory, but one aspect of it is controversial. As Cyrille Imbert 2006 has noted, Bedau's definition places no restriction on the kind of end state that can weakly emerge from a simulation except that it be nominally emergent, where a nominally emergent property is one that can be possessed at the macro level but not at the micro level. Employing a number of illuminating examples, Imbert argues that this is a defect in the definition of weak emergence and goes on to argue that the additional criteria for genuinely emergent states generated by computationally incompressible processes must be contextual and based upon the immediately preceding states in the generating process. Intuitions and arguments can pull us in different directions on this issue, and rather than argue for or against the adequacy of weak emergence as originally presented, I want to use Imbert's insights to emphasize the logical independence of weak emergence from conceptual emergence. The issue of whether to impose restrictions on the kinds of outputs generated by weakly emergent systems then turns out to involve considerations that are based on significantly different conceptions of emergence.[6] Some preliminary arguments are needed before we address the core issues involved.

---

[6] Bedau did address this issue in his 1997 article. As he wrote:

One might object that weak emergence is *too* weak to be called "emergent," either because it applies so widely or arbitrarily that it does not demark an interesting class of phenomena,

8.3.1. WEAK EMERGENCE IS ABOUT INSTANCES

In Humphreys (2008) I provided an argument that diachronic emergence concerns property instances and not types. Here is a different argument for that conclusion, formulated specifically for weak emergence. Consider a process within which weak emergence takes place over the first N time steps, the Nth state being a token of a property type P that has not previously been instantiated in the process. The rules governing the development of the process then regenerate that state at later stages in a way that is predictable in a highly compressed way—one simple way to do this is for it to next occur at step $2N$ via the rule: If P is instantiated at step m, then instantiate P at step $2m$. In such a system, the property instance at the Nth state is weakly emergent, but no further instances of P count as weakly emergent because the derivation of the $2N$th state from the Nth state is trivial. Contrast this with a system having a different set of rules which generates a state of type P after N steps and in which another token of property type P is produced by a computationally incompressible process at the $2N$th step. We can suppose that the system reverts to its initial state at step $N + 1$ and then cycles through the same sequence of states so that no states of type P occur between the Nth and $2N$th time steps. In this second case both instances of P are weakly emergent. This shows, from a specifically diachronic perspective, that, in addition to information about the state type itself, one needs information about the process leading up to a token state which generates that token in order to determine whether that token is emergent. That is, we can have two instances of the same property type, one instance of which is weakly emergent and the other of which is not.[7] This is in sharp contrast to conceptual emergence. It is property types that are conceptually emergent because definitional reducibility and irreducibility apply to predicates in isolation from their instances. (Specifically, singular features such as tropes or haecceities seem to be best treated as primitives rather than as emergent, but it may well be that emergent objects, taken here to be concrete entities denoted by uniquely referring terms, require some form of conceptual emergence for a proper representation.) For present purposes, the important point that follows from the primacy of instances is that we can have

---

or because it applies to certain phenomena that are not emergent. For example, indefinitely many, ad hoc [Game of] Life macrostates are for all we know underivable without simulation. . . . But this breadth of instances, including those that are arbitrary or uninteresting to "emergence theorists," is not a problem or flaw in weak emergence. (Bedau 1997, 395)

Bedau has confirmed in conversation that this is still his position.

[7] This is what one should expect from any element in the relational taxonomy. Talk of a property or property instance being emergent tends to suggest that emergence is an intrinsic feature of a property or its instances. Yet as soon as we keep in mind that on both the synchronic and the diachronic approaches an emergent entity must emerge *from* something, we see that this reflects the fact that emergence is unavoidably a relational property and that unless the appropriate relation holds between the property under consideration and its origins, tokens of the former can occur without their being emergent.

two instances of type-identical states within a process and both instances are
weakly emergent.

8.3.2. NOVELTY

Now consider a criterion that is often imposed on emergent features, that they
be novel. The sense of novelty that is involved in inferential accounts of emergence is different from that used in conceptual emergence and it is intertwined
with the unpredictability requirement. First, as Imbert has emphasized, what
counts as novel in the case of weak emergence has to be decided with respect
to property instances occurring within the process that leads to E's appearance. To require that a property instance be absolutely new is far too strict, for
then only the first-ever instance of a property would count as emergent. Next,
conceptual emergence always requires supplementation or replacement of the
original theoretical framework to represent the emergent state, and this sometimes takes us out of the original scientific domain entirely. Weak emergence
imposes no such condition, and a system that produces recurring instances
of the same property whose instances are produced by a computationally incompressible process is nevertheless exhibiting repeatable weakly emergent
states. The novelty consists in the impossibility of having noninductive advance knowledge of those states within the specified theoretical domain. In
terms of the Rasmussen and Barrett representation, there is no single updating
operator U' that can replace U^T.

8.3.3. RANDOM TO RANDOM

We are now in a position to assess one argument in favor of imposing a further structural condition on the definition of weak emergence. Suppose that the
system transitions from an initial random configuration to a final random configuration. If all the intermediate states are nonrandom, then the argument of
section 8.3.1 establishes that the final state is weakly emergent. But what if all
of the intermediate stats are also random? Although random patterns are token
distinct, qua random they are type identical and so no new property type has
been produced when one random array is transformed into another, and this is
something that undoubtedly underlies the hesitation in allowing that any kind of
emergence has occurred in this situation. Furthermore, it looks like none of the
successive random states are weakly emergent because the "derivation by simulation" involves only one step. But this will not be true in all cases. Synchronic
updating of cellular automata (and of agent-based models as well) rarely involves
a single transparent derivational step; what occurs in the single time step is not
only the computational updating of the state of each cell but frequently the production of a global state that is reachable only by simulation and not by traditional closed-form prediction. Finally, although these intermediate states might

seem to violate the novelty criterion for emergence, we have seen that this violation is only apparent because weak emergence can allow repeated instances of the same property as long as the computational connections between the instances are of the appropriately complex kind, as they usually will be here.

8.3.4. ORDERED TO RANDOM

Now suppose that the system makes the transition from an initially structured state to a final random state. An appeal to the motivations behind considering self-organizing systems as producers of emergent states might in this case incline us toward insisting that this should not count as a case of emergence. The basic idea behind self-organization is that large-scale structure spontaneously emerges in a dynamic system of interacting constituents (i.e., the structure emerges without the need for external interventions) solely in virtue of local interactions between those constituents, and this structure is not accidental in the sense that were the system to be restored to its initial state or to a qualitatively equivalent state and the microdynamics rerun, similar large-scale structure would, with a high degree of probability, reappear.[8] It is the apparent ability to run counter to the general trend produced by the Second Law of Thermodynamics that makes the transition from disorder to order so interesting. Order emerging from disorder strikes us as important because it is unusual; disorder arising from order as uninteresting because it is so common.

Even if one finds this argument a compelling reason to insist that random end states generated from structured states should not count as emergent, and thus that some additional constraint should be imposed on the end state of such processes, it is clear not only that the argument relies on considerations that are completely separate from those involved in the computational basis of weak emergence but that whatever criteria are developed must be relational in form to distinguish the random to random case from the ordered to random case.

8.3.5. WEAK EMERGENCE AND ONTOLOGICAL EMERGENCE

Finally, what of the connections between weak emergence and ontological emergence? Weak emergence does not entail ontological emergence for a straightforward reason, a reason that makes weak emergent different from older versions of the inferential approach. Discussions of weak emergence are often based on examples using cellular automata, and I followed that tradition above. One example in particular, the Game of Life, is a nice clean example,

---

[8] Self-organization is almost always a statistical property of a system. In many models, unusual initial conditions will prevent the usual structures from emerging.

but emergent phenomena in the Game of Life, such as gliders, are pure model phenomena; they do not, and are not intended to, represent features of systems external to the model. This reflects the special sense in which the term "simulation" is used in the definition of weak emergence—not only is it strongly skewed toward agent-based and related simulations for a discussion of other types (see Humphreys 2004, chaps. 3 and 4), but there is no requirement that the computational process is a simulation of a real system. The weak emergence approach thus bifurcates into two subgenres. One considers the possibility of predicting phenomena only within the model; the other is concerned with the possibility of predicting states that have a intended scientific interpretation. It is only the second of these that can have any connection with emergence in real systems. It is thus easier to be a purist about weak emergence as a stand-alone criterion for emergence within the first subgenre than it is within the second. When the simulation is used as a model of another system, there may then be a motivation for imposing a restriction that only states of the simulation that correspond to states of a real physical, biological, or social system can be candidates for weak emergence.

8.4. The Sparseness Issue

We can draw one further philosophical conclusion from our analysis. What I call the sparse perspective on emergence is probably the dominant perspective in the philosophical literature on emergence. The sparse perspective holds that emergence, if it exists at all, is a rare phenomenon and indeed, the sparse perspective often seems to be used as an informal touchstone for accounts of emergence—if the account makes emergence too common, that is a reason to think the account is defective. A particularly clear version of this sparse view is held by Jaegwon Kim. Having laid out the first plausible account of reduction since Ernest Nagel's version was undermined by multiple realizability arguments, Kim asserted that

> emergentism may yet be an empty doctrine. For there may not be any emergent properties, all properties being physical properties or else functionalizable and therefore reducible to physical properties. . . . So are there emergent properties? Many scientists have argued that certain "self-organizing" phenomena of organic, living systems are emergent. But it is not clear that these are emergent in our sense of nonfunctionalizability. And . . . the classic emergentists were mostly wrong in putting forward examples of chemical and biological properties as emergent. It seems to me that if anything is going to be emergent, the phenomenal properties of consciousness, or "qualia," are the most promising candidates." (Kim 1999, 19)

The sparse position on emergence seems plausible, I believe, because it tends to focus almost exclusively on synchronic emergence. If weak emergence and other forms of diachronic, computational emergence are legitimate kinds of emergence, then examples of purely diachronic weak emergence will be quite common and the sparse position correspondingly false.

8.5. Conclusion

I have argued that the three types of account in our relational taxonomy of emergence are very different from one another. This then raises the question of whether there is a higher level framework for emergence that can unify the three types. In Humphreys 1997b, I identified six features that are commonly required of emergent phenomena—novelty, qualitative difference, nominal emergence, applicability of different laws, essential interaction, and holism—and suggested that a cluster analysis based on those features might serve as a bridge between different approaches. There may well be other, better, frameworks within which all six categories of our two taxonomies fit, but the challenge now is to find and construct it.

References

Anderson, Philip W. 1972. "More Is Different." *Science* 177: 393–96.

———. 1995. "Historical Overview of the Twentieth Century Physics." In *Twentieth Century Physics*, edited by Laurie M. Brown, Abraham Pais, and Brian Pippard, 2017–32. New York: American Institute of Physics Press.

Bedau, Mark. 1997. "Weak Emergence." In *Philosophical Perspectives: Mind, Causation, and World*, edited by James E. Tomberlin, vol. 11, 375–99. Malden, MA: Blackwell.

———. 2003. "Downward Causation and the Autonomy of Weak Emergence." *Principia Revista Internacional de Epistemologica* 6: 5–50.

Broad, C. D. 1925. *The Mind and Its Place in Nature*. London: Routledge and Kegan Paul.

Castellani, Elena. 2002. "Reductionism, Emergence, and Effective Field Theories." *Studies in History and Philosophy of Modern Physics* 33: 251–67.

Darley, Vince. 1994. "Emergent Phenomena and Complexity." In *Artificial Life IV: Proceedings of the Fourth International Workshop on the Synthesis and Simulation of Living Systems*, edited by Rodney A. Brooks and Pattie Maes, 411–16. Cambridge, MA: MIT Press.

Humphreys, Paul. 1997a. "How Properties Emerge." *Philosophy of Science* 64: 1–17.

———. 1997b. "Emergence, Not Supervenience." *Philosophy of Science* 64: S337–S345.

———. 2004. *Extending Ourselves: Computational Science, Empiricism, and Scientific Method*. New York: Oxford University Press.

———. 2008. "Synchronic and Diachronic Emergence." *Minds and Machines* 18: 431–42.

Imbert, Cyrille 2006. "Why Diachronically Emergent Properties Must Also Be Salient." In *Philosophy and Complexity: Essays on Epistemology, Evolution, and Emergence*, edited by Carlos Gershenson, Diederik Aerts, and Bruce Edmonds, 99–116. Singapore: World Scientific.

Kim, Jaegwon. 1999. "Making Sense of Emergence." *Philosophical Studies* 95: 3–36.

McLaughlin, Brian. 1997. "Emergence and Supervenience." *Intellectica* 25: 25–43.

Nagel, Ernest. 1961. *The Structure of Science*. New York: Harcourt.

Rasmussen, Steen, and Chris Barrett. 1995. "Elements of a Theory of Simulation." In *European Conference on Artificial Life*, Lectures Notes in Computer Science. Berlin: Springer-Verlag.

Rueger, Alexander. 2000. "Physical Emergence, Diachronic and Synchronic." *Synthese* 124: 297–322.

van Cleve, James. 1990. "Mind-Dust or Magic? Panpsychism versus Emergentism." *Philosophical Perspectives* 4: 215–26.

Emergence

Chapters 5 through 8 chronicle attempts to articulate why synchronic approaches to emergence need to be abandoned if we are to successfully represent the full spectrum of emergent features in the world. Many of the themes developed in those chapters have now been elaborated in *Emergence: A Philosophical Account* (Humphreys 2016), and I refer the reader to that book for details.

Chapter 5 was intended to provide a solution to the exclusion problem, one version of which is described in the paper. It had other ambitions, one of which was to allow room for nonreductive ontological pluralism; not dualism in the form of a separation between the physical and the mental, but a lusher type of pluralism that allowed a range of autonomous domains within the realm of the physical itself. For dualism, where there are just the physical and mental domains, the standard levels view is that mental properties are at a higher level than physical properties or, differently, that the mental properties are dependent on physical properties. This ordering tends to be taken as natural, but the reasons for elevating the mental above the physical bear examination. The ordering is not based on compositionality criteria, because physical objects do not constitute mental objects and mental properties are not composed of physical properties. Instead, the ordering is based on asymmetric dependency relations that result from a commitment to the ontological priority of the physical. Perhaps that begs no questions, but in the diachronic approach, the temporal dependence of the mental on the physical gives us a more natural ordering because of the historical and evolutionary priority of the latter over the former. The temporal ordering of physical domains does not always follow the synchronic compositionality ordering because some synchronically fundamental particles diachronically emerged earlier than others.

The synchronic hierarchy is a natural framework to impose if one thinks of ontological reduction and ontological emergence as opposing positions, because reduction has usually been considered as reducing the subject matter from a higher level to a lower level. Under that view, the failure of one thing to reduce to another suggests that the first may stand in a relation of emergence to the second, thus inheriting the hierarchy. Not only are levels inappropriate for diachronic accounts of emergence but the synchronic position does not take proper account of the possibility that the emergent phenomena themselves may create new levels and disrupt the existing hierarchy.

Although it is reasonably clear from the arguments in chapter 5, it bears emphasizing that I was not, contrary to some commentators' claims, endorsing the levels view but adopting it for the purposes of arguing that the exclusion argument can be avoided. The exclusion argument presupposes a set of levels, and downward causation makes sense only within that hierarchical view. It should be mentioned that switching from levels to domains will not by itself eliminate the problem that "downward causation" introduces. If a domain is claimed to be causally closed, then causation from another domain to that domain will either conflict with the closure claim for the original domain or introduce overdetermination.

Although the status of mental entities is a matter for speculation given the current state of knowledge, it seems likely that qualia will turn out to be features that emerge from neural biochemical processes. When artificial brains of complexity similar to or greater than that of the human brain are constructed, consciousness of different types may be present. The senses of which we are aware are of six types if we include proprioception, and it is possible that there could be different types of consciousness and qualia that emerge not from computation but from the architecture and interactions of different physical materials.

The fusion emergence discussed in chapter 5 is subsumed in Humphreys 2016 under a more general position called "transformational emergence." Alexandre Guay and Olivier Sartenaer have developed their own, more sophisticated, form of transformational emergence in Sartenaer and Guay 2016. Santos 2015 has an antiatomistic account under the name "relational-transformational emergence" that places weight on the transformation of relations, and Ganeri 2011 also introduced an antiatomistic approach to ontology, specifically in the context of the philosophy of mind. The essence of all variants is to take transformative processes seriously as a source of radical change in ontology or in laws. My own approach uses as its focus changes in synchronically fundamental entities that have no constituents, one example being the transformation of a muon into an electron, an electron neutrino, and a muon neutrino under the influence of the weak interaction. Because muons are synchronically fundamental particles and have no constituents, this transformation cannot be construed as the decomposition of a muon into its components, and because the transformation takes place in the synchronically fundamental domain, there is no problem with downward causation. Fusion emergence is a special case of transformational emergence because in the former the original

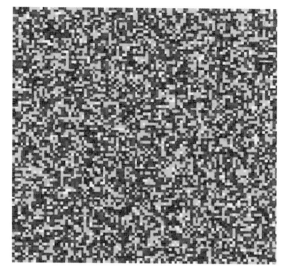

PLATE 7.2 *Initial random state of Greenberg-Hastings model*

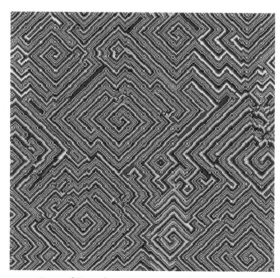

PLATE 7.3 *State of Greenberg-Hastings model after 500 time steps*

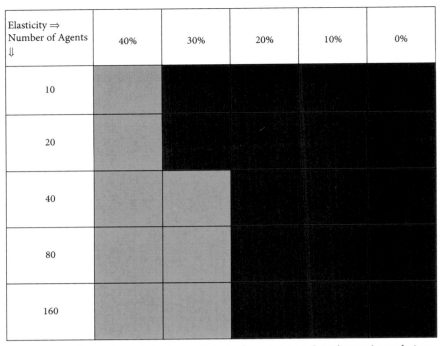

Elasticity ⇒ Number of Agents ⇓	40%	30%	20%	10%	0%
10					
20					
40					
80					
160					

PLATE 15.1A *Comparison of Maximum Gain (red) against Any Gain (orange) populations.* (*Information delay = ½, connectivity = 0.1.*)

INFORMATION DELAY

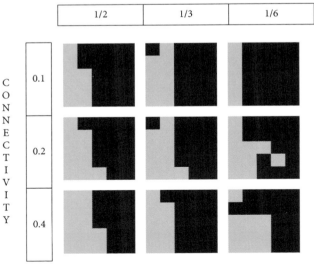

PLATE 15.1B *In red regions, Maximum Gain populations achieve superior results; in orange regions Any Gain populations are superior.*

INFORMATION DELAY

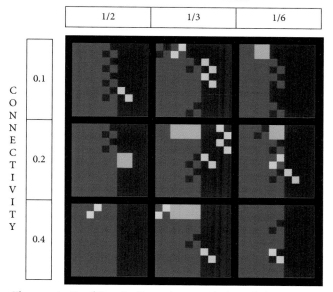

PLATE 15.2 *The superiority of Density-Dependent populations. A three-way comparison of performance of Density-Dependent (blue), Any Gain (orange), and Maximum Gain (red) search procedures for different values of connectivity and information delay. A solid color in a cell denotes the winner is a population composed of agents of that type; a checkered square indicates a two-way tie; a gray square indicates a three-way tie.*

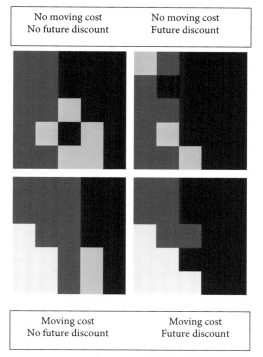

PLATE 15.3A *A four-way comparison between No Move (light blue), Density-Dependent (dark blue), Any Gain (orange), and Maximum Gain (red) populations for different welfare functions. Other axes are as in figure 15.1.*

INFORMATION DELAY

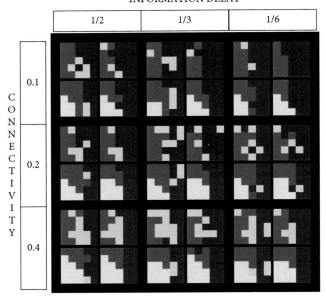

PLATE 15.3B *Four-way comparison of population performances with varying values of connectivity and information delay. Red regions are where Maximum Gain populations perform best, orange regions where Any Gain populations perform best, dark blue regions where Density Dependent populations perform best, and light blue regions where No Move populations perform best.*

entities are transformed by diachronically interacting with one another and fusing into a new noncomposite entity.[1]

Sartenaer and Guay take transformations in which a nomologically impossible situation becomes nomologically possible through the appearance of new laws. The case they examine is the fractional quantum Hall effect. Within a four-dimensional topology, the Hall conductance is present in discrete values only. When the experimental setup is confined to a three-dimensional topology (two spatial dimensions and a time dimension), plateaus of the Hall resistance appear for fractional values of the filling factor and a new type of particle called an anyon is produced, having charge values that are a fraction of the electronic charge and obeying fractional statistics. The key point that Sartenaer and Guay emphasize is that these fractional values are nomologically impossible within the original four-dimensional topology but not within the three-dimensional topology.

Chapter 6 is partly polemics, partly philosophical argumentation.[2] The overall position taken in chapter 6 is that supervenience relations are epistemologically and metaphysically too thin to provide any substantive understanding of how the supervening property depends upon the base properties. (For earlier criticisms along these lines see Horgan 1993.) It is worth restating what the overall conclusions of the article are in slightly more direct terms. The first is that, at the time this article was published, discussions of emergence had been entirely too much influenced by an emphasis on the philosophy of mind. A principal goal of the second half of the article was to argue that there is a rich source of potential examples of emergence within the physical sciences and that there are great advantages to discussing emergence using those examples as a reference point. This gain results from the fact that the physical phenomena in question are by and large well understood, in sharp contrast to the situation in the philosophy of mind in which, despite significant advances in the last fifty years, the relationship between the models and the phenomena in cognitive psychology and neuroscience is not as clear as it is in physics and in chemistry. This can lead to the view that a mere dependency relation is all that one can expect when in fact far more detailed descriptions of the relation are available in the physical sciences that are empirically well confirmed. It should surely make one pause when an apparatus that was originally designed

---

[1] Some Stoic philosophers, including Diogenes Laertius and Alexander of Aphrodisias, appear to have had some form of fusion in mind, although what is known about their theories of fusion and blending is fragmentary and therefore somewhat speculative. See Todd 1976.

[2] 1997 was something of a watershed year for philosophical articles on emergence. Aside from chapters 5 and 6, it saw the publication of Mark Bedau's "Weak Emergence" (1997), Bill Wimsatt's "Aggregativity: Reductive Heuristics for Finding Emergence" (1997), and Jaegwon Kim's "Explanation, prediction, and reduction in emergentism" (1997), the first presentation of his functional approach to reduction.

to capture the relationship between well-understood physical properties and far less well understood properties, such as aesthetic and moral properties, is applied to the relation between one set of well-understood physical properties and another well-understood physical property, where "physical" is here taken in the broad sense to include chemical, biological, geological, and other such properties. More recent attempts to capture dependency relations such as grounding are subject to similar objections. We can surely do better than that.

At the time chapter 6 was written, I put the arguments in ontological terms because of the reference in many definitions of supervenience to one property supervening on a group of other properties, a realist view that generates the ontological problem of downward causation. The most important addition I would now make to chapter 6 is to emphasize that the most plausible interpretation of supervenience in cases where multiple realizability is not present is that the supervening entity is a reconceptualization of the base properties. That is to say, in many cases when B supervenes upon A, the entity picked out by B is identical to the entity picked out by A, but it is reconceptualized in B terms rather than in A terms. This conceptual interpretation of supervenience fits well with the consistency approach advocated in chapter 6, and it also fits the influential position on emergence sketched in Anderson 1972, in which ontological reduction of the content of condensed matter physics to fundamental physics is granted, but a distinctive theoretical and methodological place is retained for the former.[3] The necessitation relation between B and A is then one of conceptual necessitation. There are no separate aesthetic properties, just ways of describing spatial arrays of color on canvas that employ terms drawn from value theory. If this conceptual interpretation is correct, then any attempt to represent emergence in terms of emergent properties supervening with nomological necessity rather than logical or conceptual necessity on base properties will fail.

Even given this conceptual interpretation, it is still possible to discuss levels because in many cases there will be ontological constitutive relations between objects that can serve as the basis for distinguishing between two levels. You are a biological object, partially composed of trillions of cells. We could assign your body to one level and individual cells to a lower level based just on that count.[4] Having made that separation, descriptions of properties specific to macroscopic features such as an overly acidic stomach would conceptually supervene at the trillion-cell level on descriptions of properties of extremely complex arrangements of cellular properties at the same trillion-cell level.[5]

---

[3] For a detailed analysis of Anderson's position, see Humphreys 2015.

[4] I am not suggesting this is the right thing to do, only that it could be done.

[5] The property of suffering from indigestion is more complicated because the psychological discomfort associated with that state involves mental states. "Complex arrangements of cellular properties" need not be restricted to mereological sums of properties.

Asserting that the properties at the trillion-cell level supervene on properties of objects at the single-cell level is either false or devoid of substantive scientific content. In some cases even this simple approach is misguided. Supposedly, molecules are composed of atoms, hence molecular properties are assigned to a higher level of objects than are properties of atoms, but this simple inclusion relation is a philosophical fiction since it does not hold for actual atoms and molecules.[6] And offloading the classification of levels on to objects cannot account for why mental properties are supposed to be at a higher level than neural properties. There are no nonneural mental objects that can instantiate the mental properties. This arbitrariness of levels in the philosophy of mind case makes the distinction between "vertical" and "horizontal" uses of supervenience moot. As a result, I would now reinterpret the rotation argument as further evidence against a property interpretation of supervenience accounts and in favor of a conceptual account couched in terms of domains rather than levels.

In section 6.3 of chapter 6 I list six characteristic features of emergent properties. Those were novelty of the emergent property; at least one qualitative difference between the emergent property and those properties from which it emerges; the logical or nomological impossibility of the emergent property occurring at a lower level; at least one different law applying to the emergent property; essential interactions playing a role in the emergence of the property; and holism. In retrospect, the third and sixth elements of that list were too influenced by the prevailing synchronic approaches to emergence. Furthermore, although the list is still useful, its elements can be derived from a smaller set of conditions. These are the fact that the emergent entity is given rise to by other features, the fact that the emergent entity is novel, and the fact that the emergent entity is autonomous. Here I have generalized the account from properties to entities so that states, objects, and perhaps some other types of entities can be emergent. The novelty condition can be put in the following way: an entity E is novel with respect to a domain D just in case it is not included in the closure of D under closure criteria C. A typical application would be to the mental and physical domains. If E is a mental state satisfying the conditions given by Brentano's thesis, D is the domain of physical states, and C is closure conditions for the physical domain, such as causal closure, that fail to generate original intentionality, then E is novel with respect to the physical. This definition of novelty subsumes the first three elements of the list given in chapter 6 if the closure conditions C are chosen appropriately, but I would revise the third condition to eliminate reference to levels and rephrase it so: An emergent property is one that could not, given the closure conditions,

---

[6] For arguments, see Humphreys 2016, 82–84.

be possessed by entities in the domain from which the property has emerged. This allows for diachronic emergence as well as synchronic emergence.

The fourth feature of emergence, that different laws apply to the emergent entity than to the entities from which they emerge, is also subsumed under the novelty condition when C includes nomological closure. The fifth property in the list, that emergent properties result from interactions between the properties of the entities included in the base domain, is of great importance to diachronic ontological emergence and also to the kinds of pattern emergence that occur in agent-based models. It is a special case of the first condition of the smaller set, that the emergent entity develops from other entities. Here interactions are dynamical features and not static relations.

As with many others, I had assumed that holism, the sixth condition, was a central, even if not a necessary, feature of emergent phenomena. I now see very clearly that imposing the holism condition is a direct consequence of synchronic modes of thought, arising within that tradition from the emphasis on compositional ontologies. There is no need to impose it on diachronic accounts of emergence because noncomposite entities can undergo diachronic emergence, although there may be holistic aspects to patterns that have emerged diachronically.

In the light of this later work on diachronic emergence, the conclusion that new values of a property are always uninteresting cases of emergence needs to be revised.[7] Under transformational emergence we can accommodate the case in which a new type of elementary particle is created by the interaction of two existing particles and the new type of particle has a value of mass that previously did not exist. The new type of particle might well count as an example of emergence even though the only distinguishing feature are new values of existing properties. This in turn means that the second condition in the original list for emergence, that the emergent entity be qualitatively different from the things from which it emerged, is not a necessary condition for emergence.

Finally, at the time I wrote chapter 6 I was unaware of Robert Batterman's 1995 article on asymptotic relations between theories. We owe a considerable debt to Batterman for his pioneering work (2000, 2002) exploring the connection between renormalization group theory and emergence. Although the grounds for some of his claims have been the subject of vigorous discussion (see, e.g., Butterfield 2011a, 2011b; Belot 2005) the fact that condensed matter physics has its own autonomous theoretical framework that cannot easily be reduced to that of high-energy physics and that there are far more sophisticated ways of dealing with multiple realizability than using supervenience

---

[7] For one version of this view see Teller 1992. Because Teller's work was some of the earliest in the modern revival of emergence, the fact that his conclusion needs to be revised is not to be taken as a criticism of his work.

relations is now widely recognized in the philosophy of science community, and one hopes that this recognition spreads to a wider philosophical world.[8] Our understanding of emergence will come through an understanding of the physical details of the processes that prevent reduction in specific cases, rather than employing an extremely general characterization of physicalism to rule out, a priori, many potential cases of emergence.

In chapter 7 the sixfold taxonomy presented is by no means the only classification that can be imposed on accounts of emergence, but I have found it useful in assessing approaches to emergence. It can be used as an overlay on finer-grained accounts of object, property, state, and nomological emergence. Epistemological emergence (Silberstein and McGeever 1999) provides an alternative classification to inferential emergence, but the two are not equivalent. I prefer the inferential terminology because it more precisely captures what I think is the core feature of the epistemic wing of the emergentist tradition, which is that the only way to first know about an inferentially emergent entity is to encounter an instance of it. This idea is present in the ideal of a complete theory for a domain of science. Call a scientific theory T complete for a domain D just in case every empirically identifiable fact about D can be represented by a sentence S in the language of T and S can be derived from T.[9] Then a necessary condition for D to contain an inferentially emergent fact is that there is no complete theory for D. The concern that it would always be possible to augment a theory T by adding to T whatever sentences are necessary to predict S is avoided because the introduction of those sentences would not be motivated until the novel fact had been encountered.

One of the unfortunate consequences of advocates of inferential emergence denying the existence of ontological emergence is that it has led to those advocates frequently not addressing the question of why a phenomenon is inferentially emergent. Inferential emergence is always relative to some inferential apparatus, but it may be that the structure of some systems prevents a complete theory from being formulated. This is one moral that can be drawn from the work of Barahona 1982 and Gu et al. 2009, in which it is the undecidability of a theory that leads to inferential emergence. This in turn raises the

---

[8] Batterman's approach to emergence is via mathematical representations and, like Butterfield's, is not directly an ontological approach to emergence. Although the appeal to infinite limits presents a challenge for a realist interpretation, Batterman's position can be seen as a contribution to the literature on mathematical explanations of physical phenomena.

[9] "Empirically identifiable" here is intended to capture the constraint that no science can empirically detect every fact in its domain, the most obvious reason being that empirical techniques are restricted to finite data sets (or countable if a modest idealization is allowed). I suspect that trying to make this constraint precise will be either very difficult or impossible—one reason to think that inferential accounts of emergence are bound by representational constraints that make inferential accounts of emergence ultimately less interesting than they initially appear to be. See the next paragraph for some arguments to that end.

difficult question of why the physical system is such that only an undecidable theory applies to it.

Discussions of reduction and emergence can take place in the ontological mode or in the theoretical mode. The philosophical literature is generally, although not universally, clearer than the scientific literature about which mode is being addressed. The scientific literature, in contrast, has a tendency to write using the ontological mode while intending to restrict itself to the theoretical mode. As mentioned above in the discussion of chapter 6, P. W. Anderson's "More Is Different" essay is a primary example of this. Although he repeatedly writes of laws and properties, he is suggesting the need to invent and employ a new linguistic and theoretical framework at each level in the scientific hierarchy. Inferential and conceptual emergence are important and philosophically interesting, but ontological emergence is the most challenging to articulate.

There are two central themes in chapter 8 that are connected. The first is that the common locution "A is emergent" misrepresents the logical form of the claim, which is properly "A is emergent from B." By ignoring this relational form, a position that is reinforced by the view that emergent phenomena are autonomous, it is easy to avoid answering the question, "Yes, but how (or why) does A emerge?" thereby conforming to the tradition that it is not possible to explain how such phenomena emerge. The second theme is that claims of diachronic emergence must recognize the essentially historical nature of this type of emergence. This is not the trivial claim that diachronic emergence includes historical features by virtue of its diachronicity. It is that one cannot tell solely by examining a current instance of a property, state, or other entity whether it is emergent or not. The argument in chapter 8 involving weak emergence demonstrates that, but the point carries over to cases of ontological emergence. This is easily seen by considering a case in which a type of entity A can be produced in two different ways, one by transformational emergence involving a new law and the other by a generative process that uses only laws that already exist. Or, to take a more speculative example, the human ability to perform moderately complicated arithmetical operations requires (in most people) conscious attention to details of the calculation. Assuming that such conscious attention is diachronically emergent from earlier evolutionary stages of development, the same ability, that of performing moderately complicated arithmetical operations, can occur in a consciousness-free computer, in which the calculations are not diachronically emergent from the computational architecture and inputs. It could be argued that those calculations are indirectly diachronically emergent in virtue of the human designers, but if so, then any humanly designed object including those arrived at by a generative atomistic process such as my brick patio will be diachronically emergent. That is an unwelcome consequence, but even if we accept it, we have, in the case of the computations, that the

same thing A can be emergent from E (conscious attention) but not from E*
(the architecture and input of an electronic computer).

References

Anderson, Philip W. 1972. "More Is Different." *Science* 177: 393–96.

Barahona, Francisco. 1982. "On the Computational Complexity of Ising Spin Glass Models." *Journal of Physics A: Mathematical and General* 15: 3241–53.

Batterman, Robert W. 1995. "Theories between Theories: Asymptotic Limiting Intertheoretic Relations." *Synthese* 103: 171–201.

———. 2000. "Multiple Realizability and Universality." *British Journal for the Philosophy of Science* 51: 115–45.

———. 2002. *The Devil in the Details: Asymptotic Reasoning in Explanation, Reduction, and Emergence.* New York: Oxford University Press.

Bedau, Mark A. 1997. "Weak Emergence." *Noûs* 31: 375–99.

Belot, Gordon. 2005. "Whose Devil? Which Details?" *Philosophy of Science* 72: 128–53.

Butterfield, Jeremy. 2011a. "Emergence, Reduction and Supervenience: A Varied Landscape." *Foundations of Physics* 41: 920–59.

———. 2011b. "Less Is Different: Emergence and Reduction Reconciled." *Foundations of Physics* 41: 1065–135.

Ganeri, Jonardon. 2011. "Emergentisms, Ancient and Modern." *Mind* 120: 671–703.

Gu, M., C. Weedbrook, Á. Perales, and M. A. Nielsen. 2009. "More Really Is Different." *Physica D: Nonlinear Phenomena* 238: 835–39.

Horgan, Terrence. 1993. "From Supervenience to Superdupervenience: Meeting the Demands of a Material World." *Mind* 102: 555–86.

Humphreys, Paul. 2015. "More Is Different . . . Sometimes: Ising Models, Emergence, and Undecidability." In *Why More Is Different*, edited by Brigitte Falkenburg and Margaret Morrison, 137–52. Berlin: Springer.

———. 2016. *Emergence: A Philosophical Account.* New York: Oxford University Press.

Kim, Jaegwon. 1997. "Explanation, Prediction, and Reduction in Emergentism." *Intellectica* 2: 45–57.

Santos, Gil C. 2015. "Ontological Emergence: How Is That Possible? Towards a New Relational Ontology." *Foundations of Science* 20: 429–46.

Sartenaer, Olivier, and Alexandre Guay. 2016. "A New Look at Emergence. Or When after Is Different." *European Journal for Philosophy of Science* 6: 297–322.

Silberstein, Michael, and John McGeever. 1999. "The Search for Ontological Emergence." *Philosophical Quarterly* 49: 201–14.

Teller, Paul. 1992. "A Contemporary Look at Emergence." In *Emergence or Reduction? Essays on the Prospects of Nonreductive Physicalism*, edited by Ansgar Beckermann, Hans Flohr, and Jaegwon Kim, 139–53. Berlin: Walter de Gruyter.

Todd, Robert B. 1976. *Alexander of Aphrodisias on Stoic Physics.* Leiden: Brill.

Wimsatt, William C. 1997. "Aggregativity: Reductive Heuristics for Finding Emergence." *Philosophy of Science* 64: S372–S384.

PART III

Probability

9

Why Propensities Cannot Be Probabilities

The notion that probability theory is the theory of chance has an immediate appeal. We may allow that there are other kinds of things to which probability can address itself, things such as degrees of rational belief and degrees of confirmation, to name only two, but if chance forms part of the world, then probability theory ought, it would seem, to be the device to deal with it. Although chance is undeniably a mysterious thing, one promising way to approach it is through the use of propensities—indeterministic dispositions possessed by systems in a particular environment, exemplified perhaps by such quite different phenomena as a radioactive atom's propensity to decay and my neighbor's propensity to shout at his wife on hot summer days. There is no generally accepted account of propensities, but whatever they are, propensities must, it is commonly held, have the properties prescribed by probability theory. My contention is that they do not and that rather than this being construed as a problem for propensities, it is to be taken as a reason for rejecting the current theory of probability as the correct theory of chance.

The first section of the paper will provide an informal version of the argument, indicating how the causal nature of propensities cannot be adequately represented by standard probability theory. In the second section a full version of the argument will be given so that the assumptions underlying the informal account can be precisely identified. The third section examines those assumptions and deals with objections that could be raised against the argument and its conclusion. The fourth and final section draws out some rather more general consequences of accepting the main argument. Those who find the first section sufficiently persuasive by itself may wish to go immediately to the final section, returning thereafter to the second and third sections as necessary.

9.1. The Informal Argument

Consider first a traditional deterministic disposition, such as the disposition for a glass window to shatter when struck by a heavy object. Given slightly idealized circumstances, the window is certain to break when hit by a rock, and this manifestation of the disposition is displayed whenever the appropriate conditions are present. Such deterministic dispositions are, however, often asymmetric. The window has no disposition to be hit by a rock when broken, and similarly, whatever disposition there is for the air temperature to go above 80°F is unaffected by whether my neighbor loses his temper, even though the converse influence is certainly there. The reason for this asymmetry is that many dispositions are intimately connected with causal relationships, and as a result they often possess the asymmetry of that latter relationship. Thus we might expect propensities, as particular kinds of dispositions, to possess a similar asymmetry and indeed they do, although because propensities come in degrees, the situation is understandably somewhat different.

The point can be illustrated by means of a simple scientific example. When light with a frequency greater than some threshold value falls on a metal plate, electrons are emitted by the photoelectric effect. Whether or not a particular electron is emitted is an indeterministic matter, and hence we can claim that there is a propensity p for an electron in the metal to be emitted, conditional upon the metal being exposed to light above the threshold frequency. Is there a corresponding propensity for the metal to be exposed to such light, conditional on an electron being emitted, and if so, what is its value? Probability theory provides an answer to this question if we identify conditional propensities with conditional probabilities. The answer is simple— calculate the inverse probability from the conditional probability. Yet it is just this answer which is incorrect for propensities and the reason is easy to see. The propensity for the metal to be exposed to radiation above the threshold frequency, conditional upon an electron being emitted, is equal to the unconditional propensity for the metal to be exposed to such radiation, because whether or not the conditioning factor occurs in this case cannot affect the propensity value for that latter event to occur. That is, with the obvious interpretation of the notation, $\Pr(R/\bar{E}) = \Pr(R/E) = \Pr(R)$. However, any use of inverse probability theorems from standard probability theory will require that $P(R/E) = P(E/R)P(R)/P(E)$ and if $P(E/R) \neq P(E)$, we shall have $P(R/E) \neq P(R)$. In this case, because of the influence of the radiation on the propensity for emission, the first inequality is true, but the lack of reverse influence makes the second inequality false for propensities. To take another example, heavy cigarette smoking increases the propensity for lung cancer, whereas the presence of (undiscovered) lung cancer has no effect on the propensity to smoke, and a similar probability calculation would give an incorrect result. Many other examples can obviously be given.

Thus a necessary condition for probability theory to provide the correct answer for conditional propensities is that any influence on the propensity which is present in one direction must also be present in the other. Yet it is just this symmetry which is lacking in most propensities. We can hence draw this conclusion from our informal argument: the properties of conditional propensities are not correctly represented by the standard theory of conditional probability; in particular any result involving inverse probabilities, including Bayes' theorem, will, except in special cases, give incorrect results.

This short argument needs refinement, and so I turn to a fuller version which has a structure similar to the one just given but which is, of necessity, somewhat more complex.

9.2. The Detailed Argument

Any standard axiomatic system for conditional probability[1] will contain this multiplication principle:

$$(\text{MP}) \quad P(AB/C) = P(A/BC)P(B/C) = P(B/AC)P(A/C) = P(BA/C)$$

I emphasize here that this relationship appears not only as a direct consequence of the traditional definition of conditional probability, viz. $P(A/B) = P(AB)/P(B)$ but also as an axiom in probability calculi which take conditional probability as a primitive relation.[2] If we assume also the additivity axiom for conditional probabilities,

(Add)   If A and B are disjoint, then $P(A \lor B/C) = P(A/C) + P(B/C)$,

then as an easy consequence we have the theorem on total probability for binary events:

$$(\text{TP}) \quad P(A/C) = P(A/BC)P(B/C) + P(A/\overline{B}C)P(\overline{B}/C)$$

and also Bayes' theorem for binary events:

$$(\text{BT}) P(B/AC) = P(A/BC)P(B/C)/[P(A/BC)P(B/C) + P(A/\overline{B}C)P(\overline{B}/C)].$$

I note here for future reference that the only additional assumption needed to derive these second two from the first two is distributivity.

---

[1] I take standard axiom systems for conditional probability to be those containing at least axioms of additivity, normalization, nonnegativity, and the multiplication principle.

[2] For example, Karl Popper, *The Logic of Scientific Discovery* (London: Hutchinson and Company, 1959), appendices *iv, *v; R. Stalnaker, "Probability and Conditionals," *Philosophy of Science* 37 (1970): 64–80, especially p. 70.

Consider now the conditional propensity function $Pr(A/B)$, the propensity for A to occur, conditional on the occurrence of B.[3] This propensity will be interpreted initially as a single-case propensity, where A and B are specific instances of event types, but nothing that is said here entails that either A or B has actually occurred or will occur. Dispositions being relatively permanent properties, they can be attributed to a system irrespective of whether the test condition, B, or the display, A, actually occurs. I shall assume throughout that the specific system which possesses the propensity remains the same, and hence omit notational devices representing the system or the structural basis of the propensity. Propensities are, however, often time-dependent, and so a fuller notation $Pr_{t_i}\left(A_{t_j}/B_{t_k}\right)$ is needed, interpreted as "the propensity at t_i for A to occur at t_j, conditional upon B occurring at t_k." I shall now show that both the multiplication principle and Bayes' theorem fail for conditional propensities. A specific example will be referred to for illustrative purposes, but the argument could be given for any case which possesses the kind of asymmetry present in the particular example. Take, then, the case of a well-known physical phenomenon, the transmission and reflection of photons from a halfsilvered mirror. A source of spontaneously emitted photons allows the particles to impinge upon the mirror, but the system is so arranged that not all the photons emitted from the source hit the mirror, and it is sufficiently isolated that only the factors explicitly mentioned here are relevant. Let I_{t_2} be the event of a photon impinging upon the mirror at time t_2, and let T_{t_3} be the event of a photon being transmitted through the mirror at time t_3 later than t_2. Now consider the single-case conditional propensity $Pr_{t_1}(\cdot/\cdot)$ where t_1 is earlier than t_2, and take these assignments of propensity values:

(1) $Pr_{t_1}\left(T_{t_3}/I_{t_2}B_{t_1}\right)=p>0$

(2) $1>Pr_{t_1}\left(I_{t_2}/B_{t_1}\right)=q>0$

(3) $Pr_{t_1}\left(T_{t_3}/\overline{I}_{t_2}B_{t_1}\right)=0,$

where, to avoid concerns about maximal specificity, each propensity is conditioned on a complete set of background conditions B_{t_1} which include the fact that a photon was emitted from the source at t_0, which is no later than t_1. The parameters p and q can have any values within the limits prescribed. We need one further assumption for the argument. It is

(CI) $Pr_{t_1}\left(I_{t_2}/T_{t_3}B_{t_1}\right)=Pr_{t_1}\left(I_{t_2}/\overline{T}_{t_3}B_{t_1}\right)=Pr_{t_1}\left(I_{t_2}/B_{t_1}\right).$

---

[3] Throughout this paper, the notation "P" will denote probability, and "Pr" propensity.

That is, the propensity for a particle to impinge upon the mirror is unaffected by whether the particle is transmitted or not. This assumption plays a crucial role in the argument, and will be discussed in the next section.

ARGUMENT 1: MP FAILS FOR PROPENSITIES

From TP we have

$$Pr_{t_1}\left(T_{t_3} / B_{t_1}\right) = Pr_{t_1}\left(T_{t_3} / I_{t_2} B_{t_1}\right) Pr_{t_1}\left(I_{t_2} / B_{t_1}\right)$$
$$+ Pr_{t_1}\left(T_{t_3} / \overline{I}_{t_2} B_{t_1}\right) Pr_{t_1}\left(\overline{I}_{t_2} / B_{t_1}\right)$$

and substituting in the values of the propensities from (1), (2), (3) above,

$$Pr_{t_1}\left(T_{t_3} / B_{t_1}\right) = pq + 0 = pq.$$

From CI we have $Pr_{t_1}\left(I_{t_2} / T_{t_3} B_{t_1}\right) = Pr_{t_1}\left(I_{t_2} / B_{t_1}\right) = q$.
Hence using MP we have

$$Pr_{t_1}\left(I_{t_2} T_{t_3} / B_{t_1}\right) = Pr_{t_1}\left(I_{t_2} / T_{t_3} B_{t_1}\right) Pr_{t_1}\left(T_{t_3} / B_{t_1}\right) = pq^2.$$

But from MP directly we have

$$Pr_{t_1}\left(I_{t_2} T_{t_3} / B_{t_1}\right) = Pr_{t_1}(T_{t_3} I_{t_2} / B_{t_1}) = Pr_{t_1}\left(T_{t_3} / I_{t_2} B_{t_1}\right) Pr_{t_1}\left(I_{t_2} / B_{t_1}\right) = pq.$$

We thus have

$$pq^2 = pq,$$

i.e., $p = 0$, $q = 0$, or $q = 1$, which is inconsistent with (1) or with (2).

ARGUMENT 2: BAYES' THEOREM FAILS FOR PROPENSITIES

Take as assumptions BT and (1), (2), (3) above. Then substituting in those values to BT we have

$$Pr_{t_1}\left(I_{t_2} / T_{t_3} B_{t_1}\right) = pq / \left[pq + 0\right] = 1.$$

But from CI we have

$$Pr_{t_1}\left(I_{t_2} / T_{t_3} B_{t_1}\right) = Pr_{t_1}\left(I_{t_2} / B_{t_1}\right) = q < 1.$$

These arguments clearly suggest that inversion theorems of the classical probability calculus are inapplicable in a straightforward way to propensities. I shall now consider some of the most important ways which might be suggested for avoiding the arguments given above.

Section 9.3. Objections, Replies, and Discussion

Objection. The argument depends crucially upon the assumption CI. Rejecting a substantial part of classical probability theory is too great a price to pay, and hence we should abandon CI.

Reply. It is clearly not enough to rely upon the intuitive plausibility of CI. That principle can, however, be justified directly in the following way. The particle has a certain propensity within the given system to impinge upon the mirror. Suppose that we were to manipulate the system's conditions so that no particle hitting the mirror was in fact transmitted, say by rendering opaque the rear of the mirror. Would that alter the propensity for the particle to impinge upon the mirror? Given what we know about such systems, it clearly would not, and we could, if desired, support that claim by showing that the relative frequency of particles impinging on the mirror was unaffected by manipulations in the conditioning factor T when all other factors were kept constant as far as possible. Similarly, were we to manipulate the conditions so that all particles hitting the mirror were transmitted, say by rendering the mirror transparent, this too would leave the propensity for impinging unaltered. Given these facts, the events T_{t_3} and $\overline{T}_{t_3}$ are irrelevant to the propensity for I_{t_2}, and they can be omitted from the factors upon which the propensity is conditioned without altering its value. Some further remarks are required here. It is essential not to impose an epistemological interpretation on CI. It is undoubtedly true that in our example transmission of the particle is *evidence for* the earlier incidence of the particle on the mirror, but we are not concerned with evidential connections, nor with any other epistemological relationships. The conditional propensity constitutes an objective relationship between two events and any increase in our information about one when we learn of the other is a completely separate matter. The tendency to interpret CI evidentially must therefore be resisted. Nor should we think of CI in terms of the relative frequencies with which one event is accompanied by another. Propensity values can, in many cases, be measured by relative frequencies, but the essence of a propensity account is that it puts primary emphasis upon the system and conditions which generate the frequencies and only secondarily upon the frequencies themselves. The issues of interest for a propensity calculus are not ones stemming from the passive observation of frequencies, but the activist ones of which frequency values remain unchanged under actual or hypothetical experimental interventions. No distinction is made within frequency interpretations of probability theory between mere associations of events and genuine causal connections, but this distinction is critical for propensities and cannot be ignored.

One final point needs to be discussed in this connection. In order to avoid having to justify the assumption CI for each case individually, we might want to refer to a general principle of the form

(CI′) If Y is causally independent of X, then $\Pr\left(Y \mathbin/ XZ\right) = \Pr\left(Y \mathbin/ Z\right)$
for all Z.

My own view is that such a general principle can be justified and used in place of the special assumption CI. To do this would, however, require a lengthy excursion into some controversial issues in probabilistic causality which are not central to the point under discussion here. In particular, it would require a general justification of a variational account of causation which is applicable to indeterministic systems. I am confident that the argument given above in favor of CI is sufficiently compelling for our present purposes, and so I shall remain with it.

Objection. The asymmetry present in the example is due to temporal asymmetry and is not therefore a property of the propensities themselves.

Reply. It is true that it is difficult to separate the asymmetry of single-case propensities from the asymmetry of temporally ordered events. However, a precisely similar argument to that of section 9.2 can be given for propensities having event types as relata, and within which no temporal ordering occurs essentially. Consider the example mentioned earlier of the neighbor who harangues his wife on hot summer days. If we let T = tirade at wife and I = intensely hot day, where now no temporal subscripts are required, and retain the propensity assignments (1), (2), and (3) of section 9.2, then it is possible to repeat the arguments of that section mutatis mutandis, and show that the multiplication principle and Bayes' theorem fail for general propensities as well. The failures thus clearly stem from the nature of propensities and not from the nature of time. This response also shows that one cannot avoid the argument by insisting that it is meaningless or inadmissible to condition upon future events. For that objection would not dispose of the argument as applied to general propensities which are not temporally dependent. Furthermore, for temporally dependent single-case propensities, given any meaningful propensity assertion under this view which is conditioned only upon earlier events, there will exist an application of Bayes' theorem, and an application of the multiplication axiom, which take that meaningful propensity assertion and transform it into a meaningless claim. Indeed, any application of Bayes' theorem to temporally ordered events will fail the meaning-preservation criterion, and the restriction of probability theory required to satisfy that criterion would eliminate use of the theorem entirely for single-case propensities.

Objection. The problem lies with the use of conditional probabilities P(B/A) to represent propensities. Instead probability conditionals of the form $P(A \to B)$ should be used. As we know,[4] the two behave differently out-side trivial cases, and so the fault lies in the mode of representation and not in the probability calculus.

Reply. This response can, I think, best be construed as a positive suggestion for an alternative approach to representing propensities. For example, some versions of causal decision theory have used the difference between condi-tional probabilities and the probability of conditionals to avoid Newcomb problems, by invoking a principle of causal independence which is similar to CI' above, so that when A has no causal influence on B, $P(A \to B) = P(B)$.[5] If such an approach is taken, however, it would have to be sharply separated from the subjectivist interpretations of the probability function with which it is usually associated, for as I construe them, propensity values are objective properties of physical and social systems.[6] Because the properties of condi-tional propensities are so intimately connected with those of probabilistic cau-sality, and there is currently available no comprehensive theory of the latter for the singular case, I am unfortunately unable at present to offer a positive account of the nature of conditional propensities.

Discussion. How do we arrive at the propensity assignments (1),(2), and (3)? Because the argument depends only upon whether the propensities do or do not have extremal values, we can invoke the following two special princi-ples both of which appear to be correct for single-case propensities, (although each would be subject to measure-theoretic nuances within a Kolmogorovian framework). The first principle is: if an instance $X_{t'}$ of an event type X never occurs with an instance Y_t of an event type Y, then the conditional propensity of $X_{t'}$ conditional on Y_t is zero, for any such pair of instances. The second prin-ciple is: if an instance of an event type X occurs together with an instance of an event type Y, and an instance of event type Y occurs without an instance

---

[4] See David Lewis, "The Probability of Conditionals and Conditional Probabilities," *Philosophical Review* 85 (1976): 297–315, in particular pp. 300–302.

[5] For example, A. Gibbard and W. Harper, "Counterfactuals and Two Kinds of Expected Utility," in *Foundations and Applications of Decision Theory*, vol. 1, ed. C. Hooker, J. Leach, and E. McClennen (Dordrecht: D. Reidel, 1978), 125–62.

[6] David Lewis, in his "A Subjectivist's Guide to Objective Chance," in *Studies in Inductive Logic and Probability*, vol. 2, ed. R. C. Jeffrey (Berkeley: University of California Press, 1980), has provided what is probably the most fully developed theory relating chance and credence. One brief point should be made in connection with Lewis's theory. For him, chance is credence objectified and (hence) chance obeys the laws of probability theory. Conditional chance is then defined in the usual manner. This entails, I believe, that such an account of chance based on subjective probabilities cannot capture the causal aspects of conditional propensities, even with the restrictions of admissible evidence imposed by Lewis. I certainly do not want to claim that the very rough sketch I have provided here of propensities is the only one possible, but it does suggest that carrying over the properties of subjective probability to chances will result in certain characteristic features of the latter being lost. A similar point can be made about the suggestion that we can define an absolute propensity measure as $c(A) = df \Pr(A/T)$, where T is any certain event, and then define a conditional probability measure in the usual way using c.

of event type X, then the propensity for $X_{t'}$ conditional on Y_t lies strictly between zero and one, for any such pair of instances. (Both principles assume that all other background factors have been conditioned into $Pr(\cdot/\cdot)$.) The first principle secures (3), the second principle secures (2), and the first half of the second principle secures (1).

Would it be possible to reject some assumption other than CI and preserve MP and BT? The only other candidates are finite additivity and distributivity (which is needed to derive TP and BT). Although there are well-known reasons for doubting the universal application of distributivity to quantum probabilities there is, I think, no good reason for supposing that it fails for propensities in general. The failure of finite additivity would be as conclusive a reason as the failure of the multiplication axiom to reject the classical probability calculus, and its failure would merely compound the difficulties for the traditional theory. However, the argument given above is so clearly directed against inversion principles that any considerations involving other parts of the calculus seem to be quite separate. The account thus ought not to be viewed as a pragmatic argument based on considerations of simplicity or convenience, but as showing directly the falsity of the multiplication principle and Bayes' theorem.

It is perhaps ironic that the first fully general version of Bayes' theorem was formulated by Laplace in order to calculate the probability of various causes which may have given rise to an observed effect.[7] Laplace was concerned with legitimizing a probabilistic version of Newtonian induction, of inferring causes from their effects, and given his deterministic views, only an epistemic interpretation of the theorem made sense for him. But when our concern is with objective chance, such inductive interests are of secondary importance, and once the metaphysical aspects of chance are separated from the epistemological, Laplace's interpretation no longer seems quite so compelling.

9.4. Consequences

What is the epistemological status of probability theory? It seems to occupy a peculiar position somewhere between the purely mathematical and the obviously scientific. The subject matter of the theory, if matter there be, has been identified with, among other things, finite class frequencies, degrees of rational belief, limiting relative frequencies, propensities, degrees of logical confirmation, and measures on abstract spaces, to name only some of the most important. This diversity of interpretations has been matched by the range of

---

[7] In Pierre Simon, Marquis de Laplace, *Theorie Analytique des Probabilities*, 3rd ed. (Paris: Courcier, 1820), 183–84.

views on the nature of the theory itself. It has been taken as a generalization of classical logic, as an abstract mathematical theory, as an empirical scientific theory, as a theory of inference perhaps distinct from but certainly underpinning the theory of statistical inference, as a theory of normative rationality, as the source of models for irregular phenomena, as an interpretative theory for certain parameters in scientific theories, as the basis for an analysis of causality, and as the reference point for definitions of randomness. Yet underlying this remarkable range of views is an equally remarkable agreement about the correct structure of the calculus itself. In particular, empiricists and rationalists may differ about the source of the probability values used in applications of the theory, but there is little disagreement about the truth of the theory—indeed, it would not be an exaggeration to say that the theory of probability is commonly regarded as though it were necessarily true.

If the arguments given in the first three sections of this paper are correct, this perception of probability theory is profoundly mistaken. It is thus worth recalling how it arose. Historically, the success of Kolmogorov's axiomatization, published in German in 1933, quickly eclipsed for scientific purposes Reichenbach's axiomatization of 1932 and the frequency theories of von Mises and of Popper, published in 1928 and 1934, respectively. Philosophically, the hegemony of standard probability theory has been reinforced by its affinities with logic. The view that probability theory is an extension of classical logic was adopted by Bolzano, Boole, Venn, Lukasiewicz, Reichenbach, Carnap, and Popper, and has been supported by results showing that, in some cases, the logical structure of the probability space can be derived from the axioms of probability theory, indicating that classical sentential logic is a special case of the structure imposed upon propositions by the theory of probability.[8] This, together with the application of the theory in a manner seemingly independent of subject matter, reinforces the conception that the theory has an epistemological status akin to that of logic. Hence one arrives at the position that the correct way to utilize probability theory within science is to first separately axiomatize a purely formal theory of probability, and nonprobabilistic axioms for specific scientific theories can then be added to this fixed set of probability axioms in exactly the same way that nonlogical axioms are standardly added to logical axioms or rules.

This approach naturally leads to the project of "providing an interpretation for probability theory" and the widespread use of the criterion of admissibility

---

[8] See, for example, Popper, *Logic of Scientific Discovery*, and H. Leblanc, "On Requirements for Conditional Probability Functions," *Journal of Symbolic Logic* 25 (1960): 238–42. It should be noted that Popper is somewhat ambiguous about the status of these results, for having asserted earlier that his calculus "is formal; that is to say, it does not assume any particular interpretation, although allowing for at least all known interpretations" (326) he then qualifies the results with "in its logical interpretation, the probability calculus is a genuine generalization of the logic of derivation" (356).

as a condition of adequacy for any interpretation of the theory. The criterion asserts that in order to be acceptable as an interpretation of the term "probability," at least within scientific contexts, the interpretation must satisfy a standard set of axioms of abstract probability theory or a close variant thereof. This approach of considering "probability" as a primitive term to be interpreted by means of an implicit definition is now so widespread as to be considered mandatory for any new account of probability, to the extent that we tend to automatically lapse into calling such accounts new interpretations rather than new theories of probability.

It is time, I believe, to give up the criterion of admissibility. We have seen that it places an unreasonable demand upon one plausible construal of propensities. Add to this the facts that limiting relative frequencies violate the axiom of countable additivity and that their probability spaces are not sigma-fields unless further constraints are added; that rational degrees of belief, according to some accounts, are not and cannot sensibly be required to be countably additive; and that there is serious doubt as to whether the traditional theory of probability is the correct account for use in quantum theory. Then the project of constraining semantics by syntax begins to look quite implausible in this area.[9] I do not wish to deny that the project of axiomatizing probability theory has had an enormously clarifying effect upon investigations into probability. What I do deny is that the concept of chance, as represented by propensities, is so obscure, or so abstract, that its properties are accessible only by means of a theory whose origins in equipossible outcomes and finite frequencies can all too easily be forgotten.[10]

---

[9] For a discussion of the flaws in the relative frequency interpretation, and a suggested repair, see B. van Fraassen, *The Scientific Image* (Oxford: Clarendon Press, 1980), chap. 6, sec. 4; on subjective probability see B. de Finetti, "On the Axiomatization of Probability Theory," in his *Probability, Induction, and Statistics* (New York: Wiley and Sons, 1972), and for the quantum-theoretical case, see A. Fine, "Probability and the Interpretation of Quantum Mechanics," *British Journal for the Philosophy of Science* 24 (1973): 1–37, for a discussion and dissenting view.

[10] This paper has benefited greatly from the comments and criticisms of Robert Almeder, James Cargile, James Fetzer, Ronald Giere, Donald Gillies, Clark Glymour, Richard Otte, Sir Karl Popper, Wesley Salmon, Robert Stalnaker, and an anonymous referee for this journal. It should not be assumed that they endorse the conclusions which I have drawn above. Brief mention of the main point can be found in my paper "Is 'Physical Randomness' Just Indeterminism in Disguise?," in *PSA 1978*, vol. 2, ed. P. Asquith and I. Hacking (East Lansing: Philosophy of Science Association, 1981), 102, and early responses are in W. Salmon, "Propensities: A Discussion Review," *Erkenntnis* 14 (1979): 213–14; and James H. Fetzer, *Scientific Knowledge: Causation, Explanation, and Corroboration* (Dordrecht: D. Reidel, 1981), 283–86.

{ 10 }

Some Considerations on Conditional Chances

10.1. Introduction

Donald Gillies's illuminating survey article on propensities (Gillies 2000) discusses a number of responses to what has been called "Humphreys' Paradox,"[1] henceforth abbreviated to HP. The essence of HP is that single-case conditional propensities can lead to an inconsistency when principles such as Bayes' theorem are used to invert those conditional propensities. There now exists a body of literature responding to HP, much of which has helped to refine our understanding of what the formal constraints on propensities must be. My purpose here is to assess the major lines of response to HP in a way that I hope will illuminate the nature of conditional propensities. My conclusion will be that none of the existing responses undermines the principal consequence of HP that conditional single-case propensities cannot be standard probabilities. This indicates that the nature of propensities cannot be properly captured by standard probability theory.

It is not often remarked that at least three different versions of the paradox have been proposed. This variety of versions has arisen because the core idea behind the problem is easily conveyed informally, and discussions which present the problem in this informal way have fastened upon different aspects of the basic problem. As a result, not all of the replies to HP are replies to the same thing. The original version of the paradox (Humphreys 1985) and the two other versions discussed here are formal paradoxes in the sense of giving rise to explicit inconsistencies. Other, more casual, formulations are paradoxes only in the sense of presenting a result which conflicts with ordinary expectations. It is the formal versions that require our attention because for those a satisfactory solution is required.

---

[1] The name originates with Fetzer 1981.

10.2. The Basic Issue

The essence of the issue can easily be conveyed. Suppose some conditional propensity exists, the propensity for D to occur conditional on C, $\Pr(D|C)$[2] For concreteness, consider the propensity of an individual to come down with influenza (event D), conditional upon his having been exposed to some other specific individual with this illness (event C). This is the kind of case to which propensities should be applicable, if single-case conditional propensities are countenanced at all, because a primary reason for introducing propensities was to deal with objective chances attached to specific situations. Standard theories of conditional probability require that when $P(D|C)$ exists, so does the inverse conditional probability $P(C|D)$. The value of $P(C|D)$ can easily be calculated within these standard theories by using the calculus of elementary probability, for example via a basic form of Bayes' theorem:

$$P(C|D) = P(D|C)\ P(C)\,/\,P(D).$$

Yet the inverse propensity in the above example, $\Pr(C|D)$, the propensity of the individual to have been exposed to the carrier, given that he gets influenza at some later time, is not related to $\Pr(D|C)$ in any simple way, if indeed it is mathematically dependent at all. One might even doubt whether such an inverse propensity exists. As we shall see in a moment, it can easily be shown that the inverse probability calculations based on Bayes' theorem and related results lead to an inconsistency when supplemented with a simple principle about conditional propensities.

10.3. The Formal Paradox

The original formal argument in Humphreys 1985 used an example as an illustration, although the example was intended to be representative of any system generating nonextremal conditional propensities. It involves photons being emitted from a source at time t and impinging upon a half-silvered mirror. Some of those photons are transmitted through the mirror, others are reflected from it, and what a particular photon[3] does is irreducibly indeterministic. We have two events $I_{t'}$ (a particular photon being incident upon the mirror) and $T_{t''}$ (that photon being transmitted through the mirror) where $T_{t''}$ occurs later

---

[2] Propensities are indicated by the notation "Pr"; probabilities by "P."

[3] For the purposes of the example I believe it is correct to speak of a particular photon. One may think of this in terms of the rate of emission of photons being so low that the probability of there being more than one emitted during a typical transit time is negligible.

than $I_{t'}$. There is also a set of background conditions B_t which includes all of the features that affect the propensity value at the initial time t, which is earlier than the times of both $I_{t'}$ and $T_{t''}$. The fact that this single-case propensity value $Pr_t(\cdot|\cdot)$ is attributed at the time t and not at the time of the conditioning event is a matter of some importance, as we shall see.

The assumptions fall into two groups. The first contains assignments of conditional propensity values:

(1) $Pr_t\left(T_{t''}\,|\,I_{t'}B_t\right) = p > 0$

(2) $1 > Pr_t\left(I_{t'}\,|\,B_t\right) = q > 0$

(3) $Pr_t\left(T_{t''}\,|\,\neg I_{t'}B_t\right) = 0$

where the arguments of the propensity functions are names designating specific physical events. They do not pick out subsets of an outcome space as in the measure-theoretic approach.[4]

The second group consists of a single principle of conditional independence. It asserts that the propensity values of earlier events do not depend upon the occurrence or nonoccurrence of later events:

$$(CI) \qquad Pr_t\left(I_{t'}\,|\,T_{t''}B_{t'}\right) = Pr_t\left(I_{t'}\,|\,\neg T_{t''}B_{t'}\right) = Pr_t\left(I_{t'}\,|\,B_{t'}\right).$$

(Note that here CI stands for conditional independence, not causal independence).

From these four assumptions alone it is straightforward to show that the use of Bayes' theorem results in an inconsistent attribution of propensity values because $Pr_t\left(I_{t'}\,|\,T_{t''}B_t\right) = 1$ when calculated from (1), (2), and (3) using Bayes' theorem, but $Pr_t\left(I_{t'}\,|\,T_{t''}B_t\right) < 1$ when calculated using CI and (2). In addition, a familiar principle of probability theory, the multiplication principle, also results in inconsistent attributions from the four assumptions together with the use of one standard feature of probability theory (the theorem on total probabilities) which does not require the use of inverse probabilities for its proof.[5]

There are three issues of philosophical interest that emerge from the replies which have been formulated to this problem. The first is that they force us to consider how we attribute numerical values to conditional propensities. Are they values to be attributed on the basis of substantive theoretical considerations, on the basis of empirical data including experimental data, or, in certain cases, on the basis of broad a priori principles? There are

---

[4] This is to keep the propensities oriented toward material, rather than formal, entities. By doing so, it requires significant metaphysical commitments to the existence of less than fully specific properties, such as "is even," in order to maintain contact with common probability attributions, the discussion of which I shall not pursue here. Those who wish to avoid such commitments may simply think in terms of their favorite theory of events.

[5] For details see Humphreys 1985, 562.

interesting differences in the attributions of such values among the published responses to the paradox, and it is revealing to isolate the principles on the basis of which these attributions are made. The second issue is the need to be quite clear about what conditional propensities are—what bears them, what are their arguments, and whether they represent degrees of causal influence. The third issue concerns the dynamics of propensities and how their values change over time.

10.4. Values of Conditional Propensities

The three forms of the paradox discussed in the literature can be characterized in terms of the principles they use to attribute values to conditional propensities $Pr_t \left(I_{t'} \mid T_{t''} \right)$ when t'' is later than t'.[6] (I assume here in addition that t is earlier than t'.) The first principle is the *conditional independence principle* CI, which claims that any event that is in the future of $I_{t'}$ leaves the propensity of $I_{t'}$ unchanged; i.e., $Pr_t \left(I_{t'} \mid T_{t''} \right) = Pr_t \left(I_{t'} \right)$. This principle reflects the idea that there exists a nonzero propensity at t for $I_{t'}$ to occur, and this propensity value is unaffected by anything that occurs later than $I_{t'}$. A second principle, which we may call the *zero influence principle*, holds that when t'' is later than t', $Pr_t \left(I_{t'} \mid T_{t''} \right) = 0$. That is, any event $T_{t''}$ that is in the future of $I_{t'}$ is such that the propensity at t for $I_{t'}$, conditional upon $T_{t''}$, is zero. This second view is appealing to those who consider conditional propensities to represent the degree of causal influence between the conditioning and the conditioned events. A third principle, which we can call the *fixity principle*, first represented in Milne 1986, claims that when t'' is later than t', $Pr_t \left(I_{t'} \mid T_{t''} \right) = 0$ or $Pr_t \left(I_{t'} \mid T_{t''} \right) = 1$. This is because by the time that the later event $T_{t''}$ has occurred, the occurrence or nonoccurrence of the earlier event $I_{t'}$ is already fixed.[7] Each of these principles might be justified on the basis of an a priori argument, on the basis of a posteriori argument based on empirical evidence, or on a case-by-case basis using experimental manipulations. We have already seen how employment of principle CI leads to an inconsistency with the use of Bayes' theorem. A parallel argument using the zero influence principle results in a similar inconsistency. In the case of the

---

[6] We can set aside the case where t' and t'' are contemporaneous to avoid quibbles. Nothing of any great importance hinges on that case, except the truth status of $Pr_t \left(I_{t'} \mid I_{t'} \right) = 1$. From here onward, I shall drop explicit notation about the background conditions B_t unless otherwise indicated.

[7] There is a fourth position which asserts that inverse propensities are in general meaningless and hence that propensities are, at best, an incomplete interpretation of the probability calculus. I shall not discuss this position because inverse propensities such as the propensity for an individual to have been exposed to a carrier given that he gets the flu at some later time are clearly legitimate objects of discussion. Terms referring to them do not lack meaning, even if the associated propensities do not exist or, what is not the same, have value zero.

fixity principle, because this is an indeterministic system, it is not determined whether the event $T_{t''}$ will occur once the event $I_{t'}$ has occurred. So there will be cases in which $\Pr_t\left(I_{t'} \mid T_{t''}\right) = 0$ on the fixity view, again producing an inconsistency with the results of Bayes' theorem. Similar arguments give rise in each case to violations of the multiplication principle, as the reader can easily check.

10.5. Interpretations of Propensities

A traditional taxonomy of propensities separates single-case accounts from long-run accounts. The former consider a propensity to be a disposition of a specific system to result in a specific outcome under specific test conditions. The latter takes a propensity to be a disposition of a specific system to produce specific values of frequencies under specific test conditions. A finer-grained account of single-case conditional propensities has emerged as a result of addressing HP. Four of these are the most important.

A *coproduction interpretation* considers the conditional propensity to be located in structural conditions present at an initial time t, with $\Pr_t\left(\cdot \mid \cdot\right)$ being a propensity at t to produce the events which serve as the two arguments of the conditional propensity. The positions of Miller (1994, 2002) and McCurdy (1996) fall into this category. Gillies (2000) has proposed a long-run version of the coproduction interpretation. Miller, who first formulated the approach in his 1994 book *Critical Rationalism*, expressed the position in a later paper in this way:

> $P_t\left(A_{t'} \mid C_{t''}\right)$ is the propensity of the world at time t to develop into a world in which A comes to pass at time t', given that it (the world at time t) develops into a world in which C comes to pass at time t''. (Miller 2002, 113)[8]

In McCurdy 1996 we find this description:

> The events described by the background conditions are responsible for the assignment of particular probability values to the members of the (previously established) event space, because these events are *responsible for the production of* the events in the event space . . . the values assigned to conditional and inverse conditional propensities are intended to provide a measure of the strength of the propensity for the system to produce the two future events in the manner specified. (McCurdy 1996, 108–9)

The key feature of coproduction interpretations is that the representations of the conditions present at the initial time t are not included in the algebra or σ-algebra of events within the probability space, in contrast to the events that

---

[8] Also in Miller 1994, 189.

occur as a result of those initial conditions, representations of which are a part of the formal probability space. The probability measure is defined only over those latter events, and the conditional probability measure is then defined in the usual way (for $P(B) > 0$) as $P(A \mid B) = P(A \& B) / P(B)$. In coproduction interpretations, the conditional propensity is attributed to the system at the initial time t and it is a propensity for the system, characterized by the background conditions B_t, to produce events in the probability space that are related by the standard definition just given. The conditional propensity is not taken to be a property of the system at the time of the later conditioning event.

A coproduction interpretation is in sharp contrast to a *temporal evolution interpretation*, which takes the propensity to have an initial value at t, with the propensity then evolving temporally, usually with its value changing under the influence of subsequent events. Within temporal evolution interpretations, the time order of events is crucial: when $t < t' < t''$, the propensity has an initial value at t, but this changes as later events occur between t and t' and the system evolves. When the conditioning event D occurs earlier than the conditioned event C (i.e., $t' < t''$), the conditional propensity $Pr_t(C_{t''} \mid D_{t'})$ is simply a temporal update of the original propensity $Pr_t(C_{t''})$ which was evaluated at the initial time t. When the time of the conditioning event D is later than that of the conditioned event C (i.e., $t' > t''$), the propensity evolves to the point where $C_{t''}$ either occurs or fails to occur at t'', and anything temporally subsequent is irrelevant. Principle CI is thus true for the temporal evolution interpretation.

Apparently similar to a temporal evolution interpretation, but in fact quite distinct, are *renormalization interpretations.* Here, instead of the temporal parameter t continuously evolving within real time, the conditioning event forces a jump from t to the time of the conditioning event, and the propensity value is determined at that new time. When $t < t' < t''$, this produces the same results as for the temporal evolution interpretation. But when t' is later than t'', the conditioning process "jumps over" the events between t and t', including t'', and determines the propensity value at time t', in contrast to the temporal evolution interpretation, where the evolution of the propensity must pass through events in their temporal order. The fixity view is entailed by the renormalization interpretation. This is because for conditional propensities such as $Pr_t(C_{t''} \mid D_{t'})$, by the time the conditions $D_{t'}$ are in place, the event $C_{t''}$ either has already occurred or has failed to occur. One caveat: in standard probability theory, the renormalization procedure allows us to identify the conditional probability $P(C \mid D)$ with an unconditional probability $P_D(C)$ restricted to a new domain of subsets of D. Because propensities are features of the world and are not properties of sets, that approach is inappropriate for propensities.

The fourth position, the *causal interpretation*, takes the conditional propensity to represent the degree of causal influence between the conditioning event and the conditioned event. An advocate of a causal interpretation, although in the context of probabilistic conditionals rather than conditional probabilities, is Fetzer (1981). The causal interpretation entails the zero influence view, on the usual position that there is no temporally backward causation.

Having laid out these positions, we can now consider the various responses to HP. I shall show that for each of the four interpretations of conditional propensities, at least one of the three principles CI, the fixity principle, or the zero influence principle, on the basis of which attributions of conditional propensity values are made, is true and hence that, for each of the four interpretations of propensities just described, a version of HP exists.

10.6. McCurdy's Response

One direct response to the original version of HP is to argue that principle CI (i.e., $\Pr_t\left(I_{t'} \mid T_{t''}B_t\right) = \Pr_t\left(I_{t'} \mid \neg T_{t''}B_{t'}\right) = \Pr_t\left(I_{t'} \mid B_t\right) = q$) is false for the photon case and, by implication, that it fails for other situations having a similar structure. This position is taken in McCurdy 1996, where, as we have seen, he defends a coproduction interpretation of conditional propensities.

I shall respond to McCurdy's arguments in two different ways. First, I shall argue that although a plausible case can be made for the coproduction interpretation in my original photon example, a structurally similar example illustrates that this response will not work in that and other cases. Second, I shall argue that the coproduction interpretation itself is seriously flawed as an interpretation of conditional propensities, at least in the sense that at best it preserves probability theory at the expense of losing the characteristic dispositional content of conditional propensities.

At the heart of McCurdy's argument against the original HP is his claim that he can establish, without any appeal to the probability calculus, that $\Pr_t\left(I_{t'} \mid T_{t''}B_t\right) = 1$, a value that is inconsistent with the assignment q < 1 given to the conditional propensity by CI together with the original assignment (2). Here is McCurdy's argument:

> Instead of utilizing the inversion theorems to determine the value of $\Pr_t\left(I_{t'} \mid T_{t''}B_t\right)$, the value can be arrived at as follows: the value of $\Pr_t\left(I_{t'} \mid T_{t''}B_t\right)$ must be one since the description of the system indicates that the system is arranged in such a manner that if the system produces a photon that is transmitted at t″ then the system must also produce a photon that impinges upon the mirror at t′. Indeed it is assignment (3) i.e. that $\Pr_t\left(T_{t''} \mid \neg I_{t'}B_t\right) = 0$ that provides the information that the system is arranged in this manner, but it is the arrangement of the photon system itself—and not the value of $\Pr_t\left(T_{t''} \mid \neg I_{t'}B_t\right)$—that demands that $\Pr_t\left(I_{t'} \mid T_{t''}B_t\right) = 1$. (McCurdy 1996, 110–11)

It should be clear from this quotation that McCurdy is not making the error of appealing to the fact that we can infer with certainty from the structure of the experimental arrangement that when a photon has been transmitted it must have been incident upon the mirror. Rather, it is the physical structure of the arrangement at t which he claims is the basis for the attribution of the value $Pr_t(I_{t'} \mid T_{t'}B_t) = 1$ and an appeal to a coproduction interpretation is clearly at work here.

The most direct evidence for this is his claim that

> The fact remains that, although the events $I_{t'}$, $T_{t''}$, and $\neg T_{t''}$ lack common causal factors between the times t' and t'', the events $I_{t'}$, $T_{t''}$, and $\neg T_{t''}$ share common causal factors that are effective between t and t'. Specifically, the photon transmission arrangement itself (described by B_t) provides a host of common causal factors. This fact is responsible for the failure of principle CI: if the system produces event $T_{t''}$, then it must have exhibited certain causal factors, some of which have an influence on the event $I_{t'}$. (McCurdy 1995, 116)

Because of some previous remarks in his paper alluding to factors responsible for the momentum of the emitted photons, it is possible that the photon example misleadingly suggests some quasi-deterministic aspects of the fundamentally indeterministic propensity at t, $Pr_t(I_{t'} \mid B_t)$. So it may help to consider a somewhat different example, consisting of a radioactive source of alpha particles and a spherical radiation detector completely surrounding the source. The detector is shielded from all other sources of radiation but is of less than perfect reliability so that not all emitted particles are detected. The propensity for an alpha particle to be emitted in a specified time period is, I hope uncontroversially, taken to be fundamentally indeterministic. Let $Pr_t(E_{t'} \mid D_{t''})$ be the propensity at time t for an alpha particle to be emitted during the short time interval t' conditional upon that alpha particle being detected at t'', where $t < t' < t''$[9]. This example is then formally identical to the photon example, and it should be clear that because of the irreducibly indeterministic nature of radioactive decay, there are no common causal factors between t and the beginning of t' (or the precise time of emission, if you prefer) on the basis of which one could truly assert that if the system produces the event $D_{t''}$ then it must have exhibited certain causal factors between t and t' some of which have an influence on $E_{t'}$. Principle CI is then true, and I believe evidently true, for this case—i.e., $Pr_t(E_{t'} \mid D_{t''}) = Pr_t(E_{t'} \mid \neg D_{t''}) = Pr_t(E_{t'})$. $Pr_t(E_{t'} \mid D_{t''})$ thus is not equal to unity as the coproduction interpretation requires. Given that this is so, HP still stands.

My second reply to McCurdy's argument is more general and applies to any coproduction interpretation of conditional propensities. A coproduction interpretation does preserve the formal structure of probabilities on the event space, but at a price. The price is that under this interpretation,

---

[9] The use of a time interval rather than an instant introduces no essentially new considerations.

the structural basis of the propensity presents the relation between the conditioning and conditioned events as a relation between probability measures rather than as a material relation between concrete events. There is no propensity relationship between the conditioning and conditioned events in the conditional probability $P(A|B)$ on this view. It is thus not a single-case conditional propensity, strictly speaking. A major appeal of single-case propensities has always been their shift in emphasis from the outcomes of trials to the physical dispositions that produce those outcomes. To represent a conditional propensity as a function of two absolute propensities, as coproduction interpretations do, is to deny that the disposition inherent in the propensity can be physically affected by a conditioning factor. This is, at root, to commit oneself to the position that there are conditional probabilities but only absolute propensities.

I now turn to the response given by David Miller in his 2002 paper to a different version of the paradox.

10.7. Miller's Response

David Miller invented the coproduction interpretation of propensities, versions of which may be found in his 1994 and 2002.[10] Miller formulates the paradox in this way:

> Intuitively we may read the term $P_t(A|C)$ as the propensity at time t for the occurrence A to be realized given that the occurrence C is realized. If C precedes A in time, this presents no extraordinary difficulty. But if C follows A, or is simultaneous (or even identical) with A, then it appears that there is no propensity for A to be realized given that C is realized; for either A has been realized already, or it has not been realized and never will be. In a late work Popper wrote that "a propensity zero means *no* propensity" (Popper 1990, p. 13). If the converse holds too, we may conclude that $P_t(A|C)$ has the value zero unless C precedes A in time. But it is easy to construct examples in which none of $P_t(A|C)$, $P_t(A)$, and $P_t(C)$ is zero. The simplest version of Bayes's theorem, to the effect that $P_t(A|C) = P_t(C|A)P_t(A)/P_t(C)$ is thereby violated. (Miller 2002, 112)

We see from this quotation that the version of HP that Miller is addressing is based on the zero influence principle, reinforced by an appeal to fixity (although not to the fixity principle as a basis for attributing propensity values.)

---

[10] In section 9.6 of his 1994, Miller constructs a novel relative frequency account of probability but its merits can be considered separately from those of propensity theories.

His response to the problem is contained in these two paragraphs:

> The objection involves a subtle misreading into the phrase 'given that' of an inappropriate temporal reference. Suppose that A is an occurrence that is realized, if it is realized at all, at time t', and that C is an occurrence that is realized, if it is realized at all, at time t". Talk of the propensity at time t for A to be realized (at time t') given that C is realized (at time t") does not mean that the realization of C at time t" is supposed to be given at time t'. It means that the realization of C at time t" is supposed to be given at time t. Of course, if t is earlier than t", then this supposition is subjunctive. But provided that t is earlier than t", there is no difficulty in principle in attributing a positive value to $P_t(A_{t'} | C_{t''})$. Note that if t" too is earlier than t', and C comes to pass at t", then there is an innocuous sense in which the occurrence of C is given at t'—by the time t' is reached, C has been realized. This is not the sense of the phrase 'given that' that is central to the theory of relative probability.
>
> Only if t is earlier than t" can $P_t(A_{t'} | C_{t''})$ differ from $P_t(A_{t'})$, the absolute propensity at t for A to be realized at t'. We may set aside as uninteresting the case in which t is not earlier than t". Now it should be obvious that to suppose at t that C comes to pass at t" is not to suppose incoherently that every occurrence dated between t and t" also comes to pass; it is not even to suppose at t that we are already at t". Provided therefore that t is earlier than t", to suppose at t that C comes to pass at t" is not to suppose either that A comes to pass at t' or that it does not come to pass at t', even if t' is earlier than t". In consequence there is, if t' is earlier than t", nothing in principle that disallows $P_t(A_{t'} | C_{t''})$ from taking any value greater than zero. Of course, the value of $P_t(A_{t'} | C_{t''})$ will be either zero or unity unless t is earlier than t". (Miller 2002, 112–13).

Miller's response to HP as he has formulated the problem is indeed a powerful response to versions of HP that are based on the zero influence principle. It is also a response to versions that are based on the fixity principle. It is not, however, a response to the original HP, because subscribing to the principle CI does not entail attributing a zero value to the relevant conditional propensity. In section 5 of Miller 2002, the coproduction interpretation of conditional propensities is once again endorsed and Miller asserts in section 1 that he is "largely in agreement with" the arguments contained in McCurdy's paper. So it is perhaps taken to be implicit that the coproduction interpretation provides an effective response to principle CI. If so, then one can bring to bear against this view the arguments presented in section 10.6 above to the effect that one has retained the structure of conditional probabilities at the expense of removing from conditional propensities what has traditionally been considered essential to them.

TABLE 10.1 Principles and interpretations of propensity values

	Principle CI	Fixity	Zero Influence
Coproduction	True	False	False
Temporal evolution	True	False	False
Renormalization	False	True	False
Causal	False	False	True

10.8. Other Possibilities

We can summarize the relation between the three principles and the coproduction interpretation of conditional propensities in the first line of table 10.1.

Our discussion of the remaining elements of the table can be brief, but it requires a little elaboration on the three other interpretations of propensities—temporal evolution interpretations, renormalization interpretations, and causal interpretations. The reader may not find all of these interpretations congenial but they do capture, perhaps somewhat crudely, attitudes toward conditional propensities that I have encountered in published and unpublished discussions of HP.[11]

8.1 TEMPORAL EVOLUTION

Recall that in this view, the value of the propensity is firmly rooted in present material conditions and that value dynamically evolves through time. In that sense, we can view the time t as a parameter that is continuously updated. Because only current and not hypothetical future situations affect the value of the propensity in this interpretation, principle CI is true. For similar reasons, the fixity principle and the zero influence principle are false.

8.2 RENORMALIZATION

Here $Pr_t\left(C_{t'} \mid D_{t''}\right)$ takes the occurrence of the event $D_{t''}$ as updating the initial probability assignment so that the conditions $D_{t''}$ at t'' are part of the basis of the propensity. With this view in mind, since t'' is later than t', the fixity principle is true and in consequence the zero influence principle will not be true in general. The principle CI will be false because $Pr_t\left(C_{t'}\right)$ will not have extremal values in indeterministic contexts.

---

[11] An issue that I want to set aside here is the effect that HP has on our theory of rational degrees of belief through Lewis's Principal Principle or something akin to it. Propensities are objective features of the world and the view that they must be constrained by subjective probabilities is not one that I find attractive. But there is no doubt that ignoring the way the world is can be financially ruinous if one is inclined to gamble.

8.3 CAUSAL INFLUENCE

Within this interpretation, the conditional propensity captures the degree of causal influence of the conditioning event D on the main event of interest C. Thus, assuming there is no temporally reversed causal influence, the zero influence principle is true for this interpretation.[12] Because the fixity principle also allows values of unity to be attributed to the conditional propensity, the fixity principle is false under this interpretation. Finally, since the degree of causal influence on $C_{t'}$ by $D_{t''}$ is zero, but in general $Pr_t(C_{t'}) \neq 0$, principle CI is also false for this interpretation.

This concludes our case-by-case evaluation of the twelve possibilities. We now see that for each of the four interpretations of conditional propensities, there exists exactly one principle governing the attribution of conditional propensities that is true under that interpretation. In consequence, for each of the four interpretations there is exactly one version of HP that shows conditional propensities cannot be probabilities. The view that propensities are probabilities cannot be saved by switching interpretations.

10.9. Propensities to Generate Frequencies

Donald Gillies (2000) adopts what he terms a "long run propensity" view. This is "one in which propensities are associated with repeatable conditions, and are regarded as propensities to produce, in a long series of repetitions of these conditions, frequencies which are approximately equal to the probabilities" (2000, 822). Although my own preference is for single-case propensities, which in certain stable conditions can ground long-run propensities through limit results, we need to consider Gillies's solution to HP, which he discusses using Milne's version and the Frisbee-producing machine example due to Earman and Salmon 1992. The latter example involves two Frisbee-producing machines, one of which produces 800 a day with 1% defective Frisbees, the other of which produces 200 a day with 2% defective. The problematical propensity is the propensity for a defective Frisbee to have been produced by the first machine. Gillies's account of this propensity, which has the value 2/3, is

> The statement $Pr(M \mid D \& S) = 2/3$ means the following. Suppose we repeat S each day, but only note those days in which the frisbee selected is defective, then, relative to these conditions there is a propensity that if they are instantiated a large number of times, M will occur, i.e. the frisbee will have been produced by machine 1, with a frequency approximately equal to 2/3. (Gillies 2000, p. 829).

---

[12] It is this interpretation that is explicitly rejected in Miller 2002, 115 and in Gillies 2000, 831.

There is in this statement a reference to only noting occurrences within which a Frisbee is defective, which tints the solution with an unnecessarily epistemic coloring, but that is easily eliminated by simply considering the set of outcomes involving defective Frisbees. With that minor adjustment, Gillies's solution has the required objectivity,[13] but it reintroduces exactly the situation from which propensity accounts were intended to rescue us—the relativization of a relative frequency to a reference class and, within von Mises's and some other frequency theories, the need to provide an objective criterion of randomness. In so doing, it loses exactly the features of propensities which proved attractive to many of us. As such, despite its ingenuity, Gillies's solution cannot be a complete account of propensities.

10.10. Conclusion

The features of propensities explored here force us to confront an important question. If conditional propensities cannot be correctly represented by standard probability theory, what does that say about the status of probability theory? In Humphreys 1985, section 4, I tentatively suggested that probability theory should be viewed as a contingent theory. David Miller suggests something rather different: "it is a factual matter whether propensities obey the calculus of probabilities" (Miller 2002, 115) and floats the idea, derived from Popper, that propensities are generalized forces. Whatever is the truth about these matters, HP is not a mere puzzle. At the very least it tells us that standard probability theory does not have the status as a universal theory of chance phenomena with which many have endowed it.

Acknowledgments

I am indebted to David Miller for correspondence and conversations about these topics and to Donald Gillies for similarly helpful correspondence. An important catalyst over the years in these discussions has been Jim Fetzer. A preliminary version of this paper was presented to audiences at All Souls College, Oxford, and the Philosophy Department at the University of Konstanz. Their reactions were instrumental in improving the arguments.

---

[13] Objective in the sense of being a feature of the objects involved.

References

Earman, John, and Wesley Salmon. 1992. "The Confirmation of Scientific Hypotheses." In Merrilee H. Salmon et al., *Introduction to Philosophy of Science*, 42–103. Englewood Cliffs, NJ: Prentice Hall.

Fetzer, James H. 1981. *Scientific Knowledge: Causation, Explanation, and Corroboration*. Dordrecht: D. Reidel.

Gillies, Donald. 2000. "Varieties of Propensity." *British Journal for the Philosophy of Science* 51: 807–35.

Humphreys, Paul. 1985. "Why Propensities Cannot Be Probabilities." *Philosophical Review* 94: 557–70.

McCurdy, C S. I. 1996. "Humphreys's Paradox and the Interpretation of Inverse Conditional Propensities." *Synthese* 108: 105–25.

Miller, David. 1994. *Critical Rationalism: A Restatement and Defence*. LaSalle: Open Court.

———. 2002. "Propensities May Satisfy Bayes' Theorem." *Proceedings of the British Academy* 113: 111–16.

Milne, Peter. 1986. "Can There Be a Realist Single-Case Interpretation of Probability?" *Erkenntnis* 25: 129–32.

Popper, Karl. 1990. *A World of Propensities*. Bristol: Thoemmes Antiquarian Books.

{ 11 }

Probability Theory and Its Models

11.1. Introduction

David Freedman's work has an unusual breadth, ranging from monographs in mathematical probability through results in theoretical statistics to matters squarely in applied statistics. Over the years, we have had many philosophically flavored discussions about each of these areas and the content of some of these discussions can be distilled into three questions. First, is probability theory a mathematical theory or does it, in virtue of its wide applicability in various areas of science, count as scientific? Second, how do we get from the abstract theory of probability to the world? Third, what, if any, are the correct interpretations of probability? Material to answer the second of these questions can be found in Freedman's many publications on statistical models, such as Freedman 2005 and Freedman 1997. Some answers to the third question are directly addressed in Freedman 1995. The answer to the first question may seem obvious to those who work in mathematical probability, and it is to Freedman—probability theory is a part of pure mathematics. Ultimately, I think that this is the correct answer but I do not think that it is quite as obvious as it might appear. In what follows I shall try to provide a unified approach to all three questions.[1]

11.2. The Status of Probability Theory

My first question is a little blunt but the founder of modern probability theory addressed it succinctly. For Kolmogorov, probability theory was to be a part of mathematics: "The author set himself the task of putting in

---

[1] Although what follows draws on Freedman's views, they should be taken as representing my own position rather than his.

their natural place, among the general notions of modern mathematics, the basic concepts of probability theory" (Kolmogorov 1933, v).[2] It is sometimes suggested that in constructing the measure-theoretic account of probability, Kolmogorov solved one half of Hilbert's sixth problem, but this is incorrect, for in the process of solving part of the problem, Kolmogorov transformed it. Hilbert's own formulation of the sixth problem makes this quite clear: "The investigations on the foundations of geometry suggest the problem: To treat in the same manner, by means of axioms, those *physical sciences* in which mathematics plays an important part; in the first rank are the theory of probabilities and mechanics. As to the axioms of the theory of probabilities, it seems to me desirable that their logical investigation should be accompanied by a rigorous and satisfactory development of the method of mean values in mathematical physics, and in particular in the kinetic theory of gases" (Hilbert 1901–2, emphasis added). This clearly indicates that for Hilbert at least, probability was viewed as a part of science, not of mathematics.[3]

As judges of what counted as mathematics the credentials of Hilbert and Kolmogorov are impeccable, and so probability theory must at some point have made the transition from science to mathematics. The task is to clearly show how empirical content is associated with probability theory while allowing it to retain its status as a part of pure mathematics. One way to reject the view that probability theory itself has empirical content is to view the measure-theoretic formulation of probability theory as a purely formal theory, as a symbolic system that has no interpretation but that imposes formal constraints on the properties of measures, random variables, and other items in the domain of the theory. If one does take this line of response, it makes answers to our second and third questions—How do we get from the formal theory to applications? and What is the correct interpretation of probability theory?—more pressing because one has to inject empirical content that is clearly probabilistic into the formal theory in order to apply it. The relation of probability theory to its applications will be somewhat different from the relation of even quite abstract scientific theories to the world, since the axioms of the latter theories already contain empirical content.[4]

---

[2] A good account of the development of modern probability theory can be found in Von Plato 1994.

[3] For a detailed examination of the relations between Kolmogorov's work and Hilbert's sixth problem see Hochkirchen 1999. I thank Michael Stoeltzner for the reference and discussions on the topic.

[4] I shall not here address a famous argument due to Quine 1951, the conclusion of which is that no sharp distinction can be drawn between mathematical and scientific theories. A response to that argument involves philosophical issues that will be addressed in a future publication.

11.3. Formal Probability Theory

To resolve this problem, and to begin to formulate answers to our three questions, we can turn to a suggestion made in Freedman and Stark 2003. There they claim: "Probability has two aspects. There is a formal mathematical theory, axiomatized by Kolmogorov [1956]. And there is an informal theory that connects the mathematics to the world, i.e., defines what 'probability' means when applied to real events" (201). This is a promising place to start, although "theory" is a little grand for the second component—something along the lines of "mapping" would be more accurate—and I would demur at the project being about meaning. The project is better construed as one addressing how probability values are correctly assigned within a model. The formal mathematical theory to which they refer is, of course, the theory developed in Kolmogorov's seminal *Grundbegriffe der Wahrscheinlichtkeitsrechnung* (1933) and later developments thereof.[5] This theory is centered on the following familiar apparatus:

Given a set Ω, a σ-algebra F on Ω and a real-valued set function P on F, the triple $<\Omega, F, P>$ constitutes a *probability space* if for any $A_i \in F$,

(1) $P(A_i) \geq 0$
(2) If $A_i \cap A_j = \emptyset$ for $i \neq j$, then $P(\bigcup_{i=1}^{\infty} A_i) = \sum_{i=1}^{\infty} P(A_i)$
(3) $P(\Omega) = 1$.

I say the standard theory is "centered" on this apparatus, because even though many discussions of how to interpret probabilities refer only to these axioms, it would be perversely narrow to identify probability theory with this minimal basis. Kolmogorov, for example, recognized the importance of the role played in his theory by the definitions of stochastic independence and conditional expectations. In addition, from the very beginning of the modern era the theory of stochastic processes has formed an essential part of the theory of probability and supplements such as ergodicity, martingales, exchangeable measures, and the like must surely also be included.

As in any formal theory, the intrinsic nature of the elements of the domain of the probability space Ω, and hence of F, is irrelevant, a fact that is captured by the use of induced probability spaces. By mapping a given probability space $< \Omega, F, P >$ within which the elements of Ω are actual outcomes (rather than formal representations of them), onto the abstract space $<\mathbb{R}^1, B^1, \mu>$ using a random variable[6] X, where B^1 is the Euclidean Borel field on

---

[5] There are other, less widely known, examples of this type such as the theories of Alfréd Rényi (Rényi 1970, 38) and Karl Popper (Popper 1959, appendices *iv and *v.) both of which take conditional probability rather than absolute probability as a primitive.

[6] Or in the more general case, sets of random variables.

$\mathbb{R}^1$ generated by the collection of intervals (a,b], $-\infty < a < b < +\infty$, and μ is given by

$$\forall B \in B^1, \mu(B) = P\{X^{-1}(B)\} = P\{\omega \mid X(\omega) \in B\}$$

The space $<\mathbb{R}^1, B^1, \mu>$ induced by the random variable X is in various generalized versions the canonical object of attention for a good deal of mathematical probability theory. The mathematical advantages provided by the induced probability space lead to the view that random variables simply "relabel" the outcomes of observations and experiments and that all of the essential probabilistic features can be found in the induced space. As a result, in standard presentations of probability theory a sharp separation is not maintained between the representations of the outcomes and the outcomes themselves, a situation which supports the formalist view and the position that if two generating systems ("trials") have the same probability space, then they are probabilistically identical.[7] This is one reason why, considered at this level of abstraction, probability theory can be considered to be a part of pure mathematics—the objects with which it is concerned are mathematical objects that can be taken sui generis.

11.4. Probability Models

Returning to Freedman and Stark's article, its immediate purpose was critical. The authors argued that predictions of earthquakes should be viewed with extreme skepticism because the models upon which the probability values were based are imprecise, poorly motivated, and based upon slim empirical evidence. I have no disagreement with that conclusion. There is also a positive message in the article that is muted yet deserves attention. It is this: rather than locating probabilities in the abstract theory or locating probabilities as features of the world, a more realistic approach is to emphasize the role of probabilistic models, and to locate the probabilistic content in those models. One significant advantage of this approach is that it makes much clearer the way in which subject-matter-specific scientific content plays a role in attributions of probability values.

This need to supplement the pure theory with specific models was recognized by Kolmogorov when he wrote in regard to his own theory: "Our system of axioms is not, however, *complete*, for in various problems in the theory of probability different fields of probability have to be examined" (Kolmogorov 1956, 3).[8] What is needed can be seen by noting that probability

---

[7] See, e.g., Itô 1984, 2.

[8] By a "field of probability," Kolmogorov meant anything that satisfies the axioms.

theory has at least two roles. There is its representational role—specifying what are the general properties possessed by probabilities—and its use as a calculus, allowing the computation of specific probability values. Until some particular measure or distribution is introduced, abstract probability theory has few nontrivial uses as a calculus. In order to keep separate these uses, we can consider the formal theory of probability as a *mathematical template* (see Humphreys 2004) within which specific models and their particular distributions can be substituted. We can usefully draw parallels between the situation in probability theory and that in classical mechanics. In the force-based version of the latter theory, certain fundamental principles are laid down that require the specification of a force function in order for the theory to be applied. Newton's Second Law $F = m\ a$ places only minimal constraints on forces and has only formal content until a particular force function has been substituted for the placeholder F. It is these fundamental principles that are the templates—they are general mathematical forms within which substitution instances can be made for purposes of empirical application. The need for this kind of substitution within highly abstract theories is common—the specification of the form of Lagrangians or Hamiltonians in other versions of classical mechanics, and the specification of basis sets and Hamiltonians in quantum mechanics are but three widely used examples of this need.

In applying the theory to particular systems, some special features of the distributions will ordinarily be used because at the level of the Kolmogorov axioms, probabilities have no internal structure beyond the minimum imposed by the axioms. From the perspective of abstract probability theory this is understandable, for many of the core results in the area are dependent upon the choice of measure or of the probability space only in very general ways, such as requiring the measure to be separable. In contrast, the specification and articulation of particular distributions is of considerable importance for applying probabilities, because the structure of the distribution is often motivated by considerations, however elementary, about the subject matter to which it applied. In what follows, I shall reserve the term "particular distribution" to denote a probability distribution having a density or mass function identifiable by a specific functional form, such as that of the hypergeometric distribution. The plain term "distribution" denotes a more abstract sense in which (in the one-dimensional case) a distribution is simply a probability measure over the Borel algebra on the reals. If the mathematical form of the substitution instance is computationally tractable the template becomes a *computational template* and the template can then be used as a calculus. In the example of classical mechanics, the templates form a familiar part of the apparatus of ordinary and partial differential equations, only some of which lend themselves to a closed-form solution, given appropriate initial or boundary conditions.

11.5. Model and Concrete Generating Systems

Here, again, is Kolmogorov: "We apply the theory of probability to the actual world of experiments in the following manner: 1) There is assumed a complex of conditions, $\mathfrak{C}$, which allows of any number of repetitions. 2) We study a definite set of events which could take place as a result of the establishment of the conditions $\mathfrak{C}$" (Kolmogorov 1956, 3). Combining this with the Freedman and Stark approach, we shall need two types of structure in addition to those making up the formal theory in order to capture the relation between probability theory and the world.

First, we have the pair <MS, MP> consisting of the *model generating system MS* and its associated *model probability distribution MP. MS* serves as the source of the elements in Ω, the outcome set of the probability space. The structure of *MS*, which can be quite abstract, constrains and sometimes even determines the structure of *MP* in a way illustrated by the Poisson model described below. *MP* is the substitution in the probability template that converts the distribution *P* occurring in the probability space into a particular distribution. Both *MS* and *MP* are mathematical objects: "probability is just a property of a mathematical model intended to describe some features of the natural world. . . . This interpretation—that probability is a property of a mathematical model and has meaning for the world only by analogy—seems the most appropriate for earthquake prediction. To apply the interpretation, one . . . interprets a number *calculated from the model* to be the probability of an earthquake in some time interval" (Freedman and Stark 2003, 5, emphasis added).

Second, we have the pair <CS, CP> consisting of a *concrete generating system CS* and the corresponding *concrete probability distribution CP*. Examples of concrete generating systems are a die thrown under specified conditions, a radioactive atom, and a stock traded on a market. These are real dice, real atoms, and real stock markets, not representations of them. Concrete systems give rise to concrete outcomes but it is very easy to conflate these concrete outcomes with our representations of them. So, to be quite clear: by the outcome of a process such as a die toss, I mean the outcome of a particular side coming up, not a representational description such as "6."

It is useful to think in terms of the hierarchy

$$< \Omega, F, P > \Rightarrow < MS, MP > \Rightarrow < CS, CP >$$

and the intermediate link is the focus of model based probabilities.[9] Everything in this hierarchy except the members of the rightmost element is a formal

---

[9] In many cases there will also be an abstract, nonmathematical, model of *CS* between *MS* and *CS* but for simplicity we can assume that object is used heuristically and is not part of the deductive apparatus.

mathematical object. In virtue of specifying the particular distribution *MP* and the relation between *MS* and Ω, one moves from the template at the left to the middle element. The relation between the middle element and the concrete system is discussed with respect to interpretations in section 11.8 below.

11.6. Some Examples

To see how the hierarchy works, consider the very simple example of a system which has the structure of a Poisson process. This example lies between the simple transparent models of coin tossing and dice throwing and the complex opaque models criticized by Freedman and Stark and so can perhaps better illuminate how model-based probability distributions are generated.

Here is one set of assumptions behind the attribution of a Poisson process to a system:

(a) During a small interval t, the chance of one event occurring in that interval is approximately proportional to the length of that interval: $P(N_t = 1) = \lambda t + f(t)$ where $f(t) \in o(t)$ i.e.,

$$\lim_{t \to 0} \frac{P\{N_t = 1\} - \lambda_t}{t} = 0.$$

(b) The chance of two or more events occurring in a small interval is small and goes to zero rapidly as the length of the interval goes to zero: $P(N_t \geq 2) \sim o(t)$.

(c) The chance of n events occurring in a given interval is independent of the chance of m events occurring in any disjoint interval, for any $n, m \geq 0$.

(d) The chance of n events occurring in an interval of length t depends only upon the length of the interval and not upon its location.

These facts about the process can be given precise mathematical representations in familiar ways and from those representations one can derive the exact form of the probability distribution covering the output from the process. Within the broad spectrum of applications of the Poisson process, there is a division between those which are justified solely in terms of a reasonable fit between the observed distribution of frequencies and the model distribution, which we can call *frequency-driven models*, and those for which some moderately plausible scientific model lies behind the adoption of the probability model, which we can call *theory-driven models*. An example of the former is the analysis of flying bomb hits on London during World War II (Clarke 1946), for which the empirical fit to a Poisson distribution is reasonably good. Yet there is no plausible aerodynamical or military model that would explain why the trajectories of V-1 rockets should satisfy this particular distribution.

In other cases, some attempt is made at providing a theory-driven model and one of the appealing features of the Poisson process is that such models can be easily generalized. The successive development of a simple model for fluctuations in electron-photon cascades in absorbers provides an illustration of this. Electron-photon cascades occur when high-energy electrons colliding with atoms in an absorber produce photons that lead to production of further electrons, producing a cascade effect. The initial model by Bhabha and Heitler published in 1937 identified *MS* with a basic Poisson process within which t represents the thickness of the absorber and n represents the number of electrons above an energy level E, so that

$$P_n(E,t) = e^{-\lambda t} \frac{(\lambda t)^n}{n!}; \; \lambda = \lambda(E).$$

However, the predicted mean value for n of λt from this particular model is physically implausible because thicker materials tend to absorb energy from electrons and so it was first modified to a linear birth process and then to a Pólya process. The fact that none of these distributions is completely realistic reflects the severe simplifications resulting from treating the energy levels as discrete and model generating systems of greater sophistication were subsequently developed.[10]

A different example of how a model generating system can be connected with a model probability distribution can be extracted from a result found in Keller (1986). Keller considered an idealized coin of negligible thickness having its center of gravity at its geometrical center (thus making it bias-free). Given the initial conditions of *u* = upward velocity of the center of gravity and ω = angular momentum of a diameter of the coin, Keller showed that when *u* → ∞ and ω → ∞ the chance of heads → 0.5, irrespective of what continuous probability density *p(u,ω)* describes the initial conditions. *MS* here consists of the mathematical representation of the idealized coin, the distribution on the initial conditions, the trajectory of the coin, and an absorbing surface. In this case, it would be appropriate to assign to a real coin a chance of coming up heads of approximately 0.5 when large values of *u* and ω are present. Note again by what thin threads the middle-level mathematical model is tied to the real system, a small amount of idealized physics sufficing to ground the model. Indeed, Diaconis, Holmes, and Montgomery (2007) question the applicability of the analysis to real coins, partly on the basis of experimental data, partly by examining the validity of the associated physical model.

---

[10] See Bharucha-Reid 1960.

11.7. Empirical Content

We now have the apparatus to answer our original question: Is probability theory a mathematical theory or a scientific theory? In the three layers of our representation

$$< \Omega, F, P > \Rightarrow < MS, MP > \Rightarrow < CS, CP >$$

the abstract probability space and the accompanying Kolmogorov theory are parts of pure mathematics that have no factual content. The development of Kolmogorov's theory may once have been partially motivated by empirical concerns, and indeed within elementary probability, relative frequencies were a guide for Kolmogorov,[11] but there is no frequency interpretation for the full nonelementary theory that is adequate.[12] In addition, the full theory makes an essential appeal to infinite collections and so has no direct empirical content. Once again, Kolmogorov: "Since the new axiom [Axiom VI, the continuity axiom, from which countable additivity follows] is essential for infinite fields of probability only, it is almost impossible to elucidate its empirical meaning. . . . Infinite fields of probability occur only as idealized models of real random processes. *We limit ourselves, arbitrarily, to only those models which satisfy Axiom VI*" (Kolmogorov 1956, 15, italics in original).

Our intermediate element, the model generating system, is also a mathematical object. Consider Freedman and Stark's example (2003, 205) of a Maxwell-Boltzmann distribution being replaced by a Bose-Einstein distribution for Bose-Einstein condensates. Initially we have a Maxwell-Boltzmann model as the middle element, the properties of which are constrained by the Kolmogorov theory.[13] Now suppose that computer-generated data from a Bose-Einstein distribution are compared with the Maxwell-Boltzmann model. They will fail to fit that model and as a result the Maxwell-Boltzmann model will be replaced by a Bose-Einstein model. Note that everything in this scenario is a mathematical object.[14] In such a case, where the form of the distribution is changed, rather than estimates of particular probabilistic parameters, it is incorrect to say that the Maxwell-Boltzmann model is revised. One constructs or selects a probability distribution for substitution in the template and that distribution either correctly represents the data or it does not. If it does not, the

---

[11] "In establishing the premises necessary for the applicability of the theory of probability to the world of actual events, the author has used, in large measure, the work of R. von Mises 1931, pp.21–27" (Kolmogorov 1933, 3 n. 4).

[12] See van Fraassen 1980, 184–87.

[13] These constraints are usually tacit and the model generating system is often considered to be an autonomous object of investigation.

[14] There is a small factual element if the data are generated on a real machine. We can either ignore that, or think in terms of a virtual machine generating the data.

process involves replacement of the model, not its revision. Now consider the situation in which data from a real Bose-Einstein system (e.g., Anderson et al. 1995) are used. Once again, if the predictions from the Maxwell-Boltzmann model do not fit the factual statistics, then one replaces the model with a better particular distribution. That particular distribution has a fixed mathematical form and any changes in its form take us to a different distribution.

Here is what we now have: There is the general mathematical template of the Kolmogorov theory and its abstract probability spaces. There is a collection of mathematical models—model generating systems. If you have Platonist inclinations, this collection is very large, it is completely abstract, and it contains models not yet known to us. If your inclinations are more constructivist, the collection will include only models known to us and built by us. All mappings between the abstract probability space and the collection of mathematical models will be mathematical. Finally, there will be a collection of real, concrete generating systems located in the world. There will be mappings between the collection of model generating systems and the collection of concrete generating systems but only a select few will be identified by users and asserted to form the basis of a modeling relation between the mathematical model and the real system. It is these mappings that contain the empirical content, not the mathematical models or the general theory. They are empirical in the sense that empirical facts about the concrete generating system play a role in whether the mapping is structure preserving. The grounds on which thee mappings are assessed is by no means simple—the problem of relating physical independence in concrete systems to stochastic independence in the model is by itself a notoriously difficult task—but the point here is that what is false is the assertion that a particular *MS* and *MP* have been correctly mapped onto a specific *CS* and *CP*. The *MS* and *MP* involved can then be replaced, but they have no empirical content themselves.

11.8. Interpretations

Finally, what about our third question concerning the interpretation of probability? In the previous section I mentioned the structural features of probability models. What of the specific probability values that are the main concern of the Freedman and Stark paper? Even from the largely formalist perspective adopted here, we cannot ignore the long tradition of trying to provide a substantive interpretation for probability theory for it underlies the differences, sometimes contentious, between subjective Bayesians, objective Bayesians, frequentists, and other schools of thought in probability and statistics. Probability theory is also tied to statistics and whether one chooses to explore the properties of loss functions or to favor classical Neyman-Pearson hypothesis testing, the interpretative issue has to be addressed. And what is perhaps most important, the present approach involves mapping a model probability distribution onto something concrete. What could that something be?

Two different approaches have been used to provide an interpretation for probability theory. The first approach uses an explicit definition of the term "probability" or "has a probability of value p." This approach was used by Hans Reichenbach (1949), Richard von Mises (1964), and Bruno de Finetti (1964). For example, in von Mises's account, as modified by Alonzo Church, we have: Event A has a probability of value p relative to the collective R if and only if A is an event of type A, A occurs in R, and events of type A occur in R with limiting relative frequency p. A collective is an infinite sequence of events within which all event types have limiting relative frequencies that are invariant under selection of subsequences by recursively definable functions on initial segments of the sequence.

Such explicit definitions have the virtue of reducing the concept of probability to other, presumably less opaque, concepts which in the case of the von Mises / Church approach are those of arithmetic limits and recursive functions and it ties the theory based on the definitions very tightly to the intended interpretation. The disadvantage is that such explicit definitions lead to accounts of probability that are different from the account provided by the standard Kolmogorov axiomatization and the theories of which are less general than the measure-theoretic account. For example, de Finetti's theory of personal probabilities, based on the concept of an agent's rational degrees of belief, rejects the countable additivity property of standard probabilities on the grounds that it is operationally meaningless. De Finetti claimed, quite plausibly, that human agents cannot distribute their degrees of belief over infinite sets of outcomes. Von Mises's frequentist theory rejected theorems of standard probability theory about events that occur infinitely often, such as the Borel-Cantelli Lemmas, on the grounds that such theorems were empirically unverifiable.

In contrast, the second approach to interpreting probability uses implicit definitions. Recognizing that chains of definitions must be grounded in primitive terms, this approach takes "probability" as a primitive and relies on a formal theory to place constraints on the probabilities. This second approach has two aspects worth noting. The first is that it captures the idea that all of the specifically probabilistic content is contained in the formal probability spaces. The second aspect is that this approach has the consequence that any probability spaces that are isomorphic are treated as indistinguishable. This is the position underlying the use of the induced probability measures discussed earlier in section 11.3. It is for this reason that the abstract theory has only formal content— any attempt to impose a more specific interpretation will be arbitrary.

The term "probability" under this second approach thus refers to an element in a formal theory. At the intermediate level there are particular formal distributions, and concrete generating systems can produce statistical estimates of values associated with those distributions. These estimates have an inescapably finite basis and can be generated using finite frequencies, rational degrees of belief, in some cases symmetry arguments, or other means. But these estimates are not interpretations of probability, they are measurements of a parameter's value. The publication of Kolmogorov's *Grundbegriffe* marked a sharp division between the formal theory of probability and those approaches, such as von Mises's and de Finetti's, that used idealizations of methods for estimating the values of elements in models, calling these idealizations "probabilities." The two sides of the division can be brought into only indirect contact.

As an analogy, consider determinism. There are formal theories of determinism that capture this intuition: a system is (historically) deterministic if, under the constraints imposed on that system by the laws that govern it, any complete state of the system is mapped onto a unique later state of the system.[15] There are model systems that are deterministic according to these theories of determinism.[16] And there are concrete systems that, within measurement error, behave similarly to the model systems. So to ask, "Are there systems that are deterministic?" is a sensible question with an affirmative answer. But the question, "What is determinism?" when asked within this theory is misplaced. All we have is the formal abstract theory together with some specific deterministic systems. And so too with probability. There is the abstract formal theory and there are the various particular probabilistic models. The question, "What is probability?" is properly approached through the latter, not the former.

How does factual probabilistic content find its way into the models? There will occur within the model generating system some parameter or distribution representing probabilities, the fact that they are probabilities being grounded in their satisfying the constraints placed on them by the mathematical theory. With frequency-driven models, the probabilities will be interpreted as frequencies; with theory-driven models, theoretically grounded input will help constrain probability values and distributional forms; perhaps even Bayesian methods can be brought to bear on other types of models. This, I believe, is where the Freedman and Stark model-based probability view constitutes both a distinctive position and a practical but cautionary note. It is distinctive because it directs the interpretative enterprise away from the theory of probability and toward specific probability models. It is practical because it forces one to consider what subject-matter-specific knowledge is required and is available to inject values into those models. And it is cautionary because it draws our attention to the fact that in many, perhaps most, models the amount

---

[15] See Montague 1962.

[16] But considerably fewer than the famous claim in Laplace (1825) (1995) suggested.

of information available is far less than we need to make serious numerical assignments of probability.

Acknowledgments

Thanks to David Freedman for his many comments on drafts of this paper. He continues to disagree with some of the claims made here.

References

Anderson, M. H., J. R. Ensher, M. R. Matthews, C. E. Wieman, and E. A. Cornell. 1995. "Observation of Bose-Einstein Condensation in a Dilute Atomic Vapor." *Science* 269: 198–201.

Bharucha-Reid, A. T. 1960. *Elements of the Theory of Markov Processes and Their Applications.* New York: McGraw-Hill.

Clarke, R. D. 1946. "An Application of the Poisson Distribution." *Journal of the Institute of Actuaries* 72: 48.

De Finetti, Bruno. 1964. "Foresight: Its Logical Laws, Its Subjective Sources." In *Studies in Subjective Probability*, edited by Henry E. Kyburg and Howard E. Smokler, 93–158. New York: Wiley and Sons.

Diaconis, Persi, Susan Holmes, and Richard Montgomery. 2007. "Dynamical Bias in the Coin Toss." *SIAM Review* 49: 211–35.

Freedman, David. 1995. "Some Issues in the Foundations of Statistics." With comments by James Berger, E. L. Lehmann, Paul Holland, Clifford Clogg, and Neil Henry, with a rejoinder. *Foundations of Science* 1: 19–83.

———. 1997. "From Association to Causation via Regression." In *Causality in Crisis?*, edited by V. McKim and S. Turner, 113–82, with discussion. South Bend, IN: University of Notre Dame Press.

———. 2005. *Statistical Models: Theory and Practice.* Cambridge: Cambridge University Press.

Freedman, David, and Philip Stark. 2003. "What Is the Probability of an Earthquake?" In *Earthquake Science and Seismic Risk Reduction*, edited by Franceso Mulargia and Robert J. Geller, 201–13. Dordrecht: Kluwer Academic Publishers.

Hilbert, David. 1901–2. "Mathematical Problems." *Bulletin of the American Mathematical Society* 8: 437–79.

Hochkirchen, Thomas. 1999. "Die Axiomatisierung der Wahrscheinlichkeitsrechnung und ihre Kontext." In *Von Hilberts sechstem Problem zu Kolmogoroffs Grundbegriffen.* Göttingen: Vandenhoeck & Ruprecht.

Humphreys, Paul. 2004. *Extending Ourselves: Computational Science, Empiricism, and Scientific Method.* New York: Oxford University Press.

Itô, Kiyoshi. 1984. *Introduction to Probability Theory.* Cambridge: Cambridge University Press.

Keller, Joseph B. 1986. "The Probability of Heads." *American Mathematical Monthly* 93: 191–97.

Kolmogorov, Andrei N. (1933) 1956. *Foundations of the Theory of Probability.* 2nd English ed. New York: Chelsea. Translation of *Grundbegriffe der Wahrscheinlichkeitsrechnung* (Berlin: J. Springer).

Laplace, Pierre-Simon (1825) 1995. *Philosophical Essay on Probabilities.* Translated from the fifth French edition of 1825 by Andrew I. Dale. New York: Springer.

Montague, Richard. 1962. "Deterministic Theories." *Decisions, Values and Groups* 2: 325–70. Reprinted as chapter 11 of *Formal Philosophy: Selected Papers of Richard Montague,* edited by Richmond H. Thomason (New Haven: Yale University Press, 1974).

Popper, Karl. 1959. *The Logic of Scientific Discovery.* London: Hutchinson.

Quine, W. V. O. 1951. "Two Dogmas of Empiricism." *Philosophical Review* 60: 20–43.

Reichenbach, Hans. 1949. *The Theory of Probability.* Berkeley: University of California Press.

Rényi, Alfréd. 1970. *Foundations of Probability.* San Francisco: Holden-Day.

van Fraassen, Bas. 1980. *The Scientific Image.* Oxford: Oxford University Press.

von Mises, Richard. 1931. *Wahrscheinlichkeitsrechnung und ihre Anwendung in der Statistik und theoretischen Physik.* Leipzig: Deuticke.

———. 1964. *The Mathematical Theory of Probability and Statistics.* Edited by Hilda Geiringer. New York: Academic Press.

Von Plato, Jan. 1994. *Creating Modern Probability.* Cambridge: Cambridge University Press.

Probability and Propensities

Each of the three articles in the section is driven by the same underlying question: is probability theory a part of mathematics or is it a very general scientific theory? During my graduate studies, I had developed a serious interest in both Andrei Kolmogorov's measure-theoretic formulation of probability theory and Richard von Mises's formal theory of *Kollectivs*.[1] Von Mises was a resolute positivist, and in his posthumous treatise on probability (1964) he rejects some central theorems of Kolmogorov's theory on the grounds that they require infinitary operations.[2] Von Mises's theory is just one of a number of variant theories of probability; we must sharpen our question. Given the variety of subject matters that have been called "probability," it is unsurprising that there is no unique theory of probability. Rational degrees of belief and logical probability are subject to normative constraints; relative frequencies, whether finite or infinite, are mathematical objects; physical chances seem to require an empirical treatment. So the sharpened question should be, "Is mainstream probability theory, represented by the standard Kolmogorovian theory, a part of mathematics, or is it a very general scientific theory in the sense that it could be shown false by some empirical phenomenon?"

The way in which such questions are formulated is important. Many philosophers have been convinced by Quine's arguments (1951) that the dichotomy underlying my question is a false one and that it is not possible to provide a sharp epistemological division between mathematical theories and scientific theories. Quine had approached the distinction using the revisability

---

[1] The connection was an interest in randomness via the then-new theory of Kolmogorov complexity for finite sequences, a topic to which Gregory Chaitin and Ray Solomonoff also made significant contributions. Although philosophers know of Kolmogorov mostly through his contributions to probability, the scope of his mathematical creativity is extraordinary. For a brief assessment see Humphreys 2007.

[2] Although von Mises's popular book *Probability, Statistics, and Truth* (1939) is better known and is more overtly philosophical in tone, the effects of his positivism on his mathematics are best seen in his 1964 book.

of theories as the central concept and argued that mathematical theories could, in extremis, be revised in the light of new empirical evidence. This approach is questionable. Mathematical theories, qua pure mathematics, are not revised. Mathematicians may lose interest in them, they may turn out to be unsuitable for particular applications, but their identity is fixed. The theory of quaternions, developed in the nineteenth century for applications in mechanics, was largely replaced by vector approaches as the preferred representation in that area of physics. As a theory of pure mathematics, it was not revised even though it had been found to be unsuitable for its intended application, and in fact the theory underwent a minor revival in the twentieth century as a representation for spatial rotations. A similar attitude should be adopted toward Kolmogorov's version of probability theory. Whether or not it eventually turns out to have too narrow (or too wide) a scope for representing chancy phenomena, the identity of that theory via its axiomatic formulation will remain fixed, it is part of pure mathematics, and, as we shall see, it is not the correct theory for conditional propensities.

PIII.1. Probability Theory

The status of Kolmogorov's theory is addressed in chapter 11, which was written for a Festschrift in honor of David Freedman, a statistician who made important contributions to both theoretical and applied statistics. The paper was aimed at a mathematical audience, and its philosophical content is somewhat muted. So I shall take this opportunity to amplify the philosophical points a little.

I take the formal theory of probability to be the modern measure-theoretic theory. The identification of probability theory with that formulation might be criticized on the grounds that probability is used with various degrees of informality in different fields and there are important philosophical issues that can only be addressed through those more informal treatments. Moreover, there are other formulations of probability theory, not all equivalent to the Kolmogorov theory, such as those of sub- and superadditive probabilities (Walley 1991; Reichenbach 1949; von Mises 1964; Rényi 1970; and others). But the full power and range of scientific applications of probability theory is gained through the abstract theory described in chapter 11. The theory of probability is not a purely formal theory, where a purely formal theory is one within which the constituent symbols have no interpretation, as I misleadingly suggested at various places in chapter 11. This is because at least the parts of the theory that employ standard mathematical terms such as Lebesgue integrals, infinite summations, and the null set already have a mathematical interpretation.[3] It

---

[3] In so saying, this begs the question neither in the direction of Quine's position nor its denial. We can rely on a widely recognized distinction between mathematical theories and scientific theories without taking a position on whether the former do or do not have empirical content.

is possible to start with a purely formal analogue of the mathematical theory, but to do so would involve complications that are an unnecessary distraction.

The content of the mathematical theory of probability involves far more than the three axioms of elementary probability theory. Generally speaking, it is specific probabilistic models that are needed to apply that abstract theory to real systems. Kolmogorov's mathematical theory is a schema, what I have elsewhere (Humphreys 2018) called a formal template, that needs to be supplemented by models of specific empirical phenomena in order to be applied to nonmathematical systems. The content of the mathematical theory of probability thus involves far more than the three axioms of elementary probability theory and a definition of conditional probability. Nevertheless, the familiar axioms and definition place clear constraints on every probabilistic model, and those constraints may not be appropriate for certain empirical phenomena.

Metaphors are always a dangerous tool in philosophy, but if one is inclined toward them, I suggest replacing Quine's famous web metaphor with a different image. Kolmogorov's theory is like the handle of a socket set on to which different sockets—specific models—are attached and detached. Neither the theory nor the models are revised when a particular model fails to fit. One just chooses a different socket, or in extremis a different handle, when needed.

The essential element in the approach is contained in section 11.7 of chapter 11. There I make the claim that all of the empirical content of the modeling process is contained in the mappings between the mathematical model and the concrete system. It is because of this that we can preserve both the abstract model and the overarching probability theory as purely mathematical objects. Moreover, not everything in the mathematical model is mapped onto parts of the concrete system. Many parts, including idealizations, will have no corresponding element in the world. For a more detailed account of this claim about empirical mappings, see Humphreys 2018, section 5.

It is widely, and venerably, held that mathematical theories are necessarily true, with the necessary truth being interpreted as truth in all possible worlds. One hesitates to challenge this view, which has much to commend it, but it can be misleading. That arithmetic is true does not mean that arithmetic applies correctly to all concrete systems. For the usual arithmetical addition operations to correctly apply, such as in $2 + 2 = 4$, the entities that are represented by the first and second tokens of 2 must retain both their identity and their individuation conditions across application of the addition operation. And elementary arithmetic does not apply to all concrete entities because those conditions are not always satisfied. It does not apply when the addition operation is interpreted as a diachronic concatenation operation applied to clouds, for example.

PIII.2. Propensities

The overall conclusion of chapters 9 and 10 is that, when supplemented by the principle CI, the standard theory of probability does not provide a correct representation for many propensities that occur in natural systems. I take propensities to be properties of natural, social, or artificial systems in which chance outcomes occur. Chance outcomes occur only in indeterministic systems, those for which a given complete instantaneous state, perhaps together with the set of all prior states, is compatible with more than one complete state for the system at some subsequent time t. One feature of propensities that is of great importance is the fact that, uniquely among objective accounts of chancy phenomena, propensities hold out the promise of allowing us to understand single-case chances in a way that provides a link to serious scientific theories. If one allows that single-case propensities occur in indeterministic systems, and that a theory of probability should correctly represent such systems, and that principle CI is true of at least some of those systems for reasons independent of the theory of probability, then the overall conclusion of chapter 9 follows: that the standard theory of probability is incorrect for many conditional propensities. It does not follow from this conclusion that the standard theory of probability is empirical, or that is has empirical content, or that it is a scientific theory.

There have been many ingenious attempts to avoid the paradox. Here I shall focus on one strategy that seems promising. It originates with Donald Gillies's suggestion (2000, chaps. 6, 7) that we add to standard probability spaces <Ω, I, P> a fourth element S which represents a set of repeatable conditions. This separation between the generating conditions for the outcomes and the events in the probability space is important because those generating conditions are not represented in the events contained in I, even though the outcomes from S are elements of Ω. Those events can only be sets constructed from elements in Ω. Furthermore, propensities are often time dependent, and in the examples considered in chapters 9 and 10, the chance setup that generates the propensities is temporally updated at various points. If the temporal updating is done only by probabilistic conditionalization on events in I, it will not properly capture the dynamics of the chance setup because those involve S. However, Gillies's goal is to preserve the Kolmogorov axioms within his theory of propensities, and his position requires accepting a long-run account of propensities rather than a single-case account. Valuable as Gillies's approach is, to abandon the single-case version of propensities is to give up one of the primary benefits of a propensity account. So it is worth considering an alternative approach presented in terms of theoretical templates and models. We can ask whether there is some formal framework analogous to Kolmogorov's theory of probability that is true for every substitution of a propensity function for the propensity placeholder. Switching to an alternative axiomatization of probability theory as it is standardly conceived will not necessarily avoid the paradox, for Aidan Lyon (2014) has shown that using Alfréd Rényi's theory of conditional probability, within which conditional probabilities are primitive rather than defined, leads to similar paradoxical consequences. However, the axiomatic formulation of a theory of propensities given in Ballentine 2016 is not equivalent to Kolmogorov's theory of probability, and Ballentine shows that the weak law of large numbers, which gives a connection between propensities and relative frequencies, can be derived from those axioms and that Bayes' theorem cannot. The axiomatization is accompanied by two crucial moves for avoiding the paradox. The first is to recognize that the temporal updating of propensity values needs to be correctly represented. The second is to separate the event space of propensities from the states of the system. These are similar to options developed in Gillies 2000.

The essential idea can be represented within the general framework of theoretical templates, models, and empirical mappings that was mentioned earlier. First, take a formal theory of propensities, such as Ballentine's, within which it is known that the paradox cannot be constructed. Then replace the second-order propensity function with a formal model for the dynamics of the propensity within a specific system. That move is needed because the dynamics of the chance setup require a model that takes into account specific features of the physical system, and those dynamics are not fully captured by probabilistic conditionalization involving outcomes. Put in slightly more technical terms, the probability space is not taken to be a fixed object but part of a system-specific model subject to constraints that are not part of probability theory proper. Finally, map the formal model on to a concrete system. Because of this use of system-specific models, propensities are different from relative frequency accounts, axiomatic accounts, and rationality-based accounts of probability in that it is unlikely that there will be a general theory of propensities.

Unlike Gillies, Mauricio Suárez has defended a single-case approach and insists that propensities should not be viewed as an interpretation of probability but should be represented by indexed chance functions, where the generating conditions for the outcomes serve as the index rather than being eligible as conditioning factors in a conditional probability function. (For details see Suárez, forthcoming.) Suárez (2013, 2014, forthcoming) has also argued that an identity thesis between conditional propensities and conditional probabilities should be rejected in both directions. The identity thesis has two parts. The claim that all propensities are conditional probabilities is the "propensity to probability thesis." It is this that the paradox shows to be false when conditional probabilities are represented by the standard axioms for probability. The converse thesis that all conditional probabilities are propensities is clearly indefensible, even if one thought that rational degrees of belief are based on propensities.

The goal of chapter 10 was to explore in a comprehensive way what were at that time the principal attempts to escape from the original paradox and to show that in approaches that were successful in that endeavor, related formal paradoxes could be constructed. Some of David Miller's criticisms (2002) concern what I called the "coproduction" interpretation of propensities. I find that interpretation quite clear, but since others do not, let me say some more about it. I wrote, "A co-production interpretation considers the conditional propensity to be located in structural conditions present at an initial time t, with $Pr_t(.|.)$ being a propensity at t to produce the events which serve as the two arguments of the conditional propensity" (Humphreys 2004, 671). Note again here the distinction that is made between events and structural conditions. The use of the term "structural conditions" was designed to capture a key feature of propensities, which is that propensities are grounded in, or arise from, what Hacking called "chance setups," that is, systems possessing structural properties that in some cases, when changed, alter the values of the associated propensities.

As an illustration, consider an individual playing Russian roulette. The structure of the system consisting of the revolver and the individual at t can produce the event $C_{t'}$ of a cartridge being in front of the firing pin at $t' > t$ and also the event $D_{t''}$ of the cartridge firing at $t'' > t'$. The value of the conditional propensity $Pr_t(D_{t''}|C_{t'})$ for the cartridge firing at t'' given that the cartridge is in front of the firing pin at t' is dependent on the cited conditions at t. When I wrote, "The key feature of co-production interpretations is that the representations of the conditions present at the initial time t are not included in the algebra or sigma-algebra of events within the probability space, in contrast to the events that occur as a result of those initial conditions, representations of which are a part of the formal probability space" (Humphreys 2004, 672), the point was that it was built into the formalism that the background conditions could not be represented in probability theory itself and that this formal deficit made the use of the usual definition of conditional probability the only available option.

The aspect of the coproduction interpretation that matters for conditional propensities is most clearly articulated by McCurdy: "The values assigned to conditional and inverse conditional propensities are intended to provide a measure of the strength of the propensity for the system to produce the two future events in the manner specified" (1996, 108–9). One of the principal differences between Drouet 2011 and Miller is that the former suggests that we can avoid the paradoxes if we consider only absolute propensities, whereas Miller holds that it is conditional propensities that are primary. Given Miller's statement of his own position, I am perfectly happy to abandon the claim in Humphreys 2004, which was based on some statements in his 2002 paper, including the claim that he is "largely in agreement with" the arguments in McCurdy's 1996 paper. However, I do not understand Miller's complaint that my assessment of possibilities regarding the principle CI are incoherent. What I wrote may be unclear, but it is not incoherent (or worse). I argued that under the coproduction interpretation offered by McCurdy, principle CI was true for conditional single-case propensities in the examples I had offered. When CI is true, the unwelcome consequences of the paradox hold. I then went on to consider another way of understanding what an advocate of a coproduction interpretation might be committed to, which is that in virtue of defining conditional probabilities in the usual way, a coproduction approach is no longer describing a single-case conditional propensity account (Humphreys 2004, 675).

Drouet 2011 is correct to point out that the determinants of probability functions are different from the arguments of probability functions themselves. It is worth noting that although within his own theory Richard von Mises was very clear about the need to specify determinants for *Kollectivs* (and for so-called "absolute probabilities" as well), the objects of interest for relative frequency accounts are arithmetical objects (sequences of outcomes) rather than concrete systems. Furthermore, because of the abstract nature of measure theory, the issue of determinants was of no real interest to Kolmogorov, despite some passing remarks about generating systems. If one wants to develop an account of updating for absolute propensities, then one needs to have a scientific theory about the dynamics of the physical determinants. What determines an appropriate measure is the physical structure of the system. I would not follow Drouet in using similarity relations within a propensity interpretation to capture changes in the system because similarity relations are too closely associated with a lack of objectivity, requiring a priori analyses to assess degrees of similarity.

The conditional propensity notation leads to an ambiguity as to what is being represented. If one emphasizes the temporal subscript t in $Pr_t(D_{t''}|C_{t'})$, then what is being represented is the propensity at t to produce the outcome $D_{t''}$ with a trajectory through state space that passes through the state including $C_{t'}$. This emphasis on the initial time prevents either of what, in chapter 10, I called the temporal evolution approach or the renormalization approach for conditional propensities from being applied. For example, what was represented in the alpha particle example of chapter 10 by $Pr_t(E_{t'}|D_{t''})$ (the propensity at t for an alpha particle to be emitted at t', conditional upon that particle being detected at t'', where t < t' < t'') is represented as the propensity at t to initiate a process that includes states corresponding to $E_{t'}$ and $D_{t''}$. Because the propensity is being evaluated at t, there is no temporal updating of the propensity to t' or t'' and in particular the fixity interpretation is inapplicable. In contrast, if what is being represented is the temporal updating to the state represented by the conditioning event, then one must evaluate the propensity at the time of the conditioning event. Thus, in $\Pr_t(D_{t''}|C_{t'})$ the absolute propensity for $D_{t''}$ is updated to t', so that if $t' < t''$, the absolute propensity $\Pr_{t'}(D_{t''})$ can be less than 1, whereas if $t'' < t'$, $\Pr_{t'}(D_{t''})$ = either 0 or 1 depending upon what happened in the real system. This indicates that the temporal evolution approach is primary in absolute propensities.

What does this mean for the admissibility of absolute propensities construed in this way?[4] If we interpret $\varnothing$ as the null state (i.e., the physical state in which there is nothing rather than one in which nothing changes), then $\Pr(\varnothing) = 0$ is true except, perhaps, at the last instant of a temporally finite universe.[5] Normalization, $\Pr(\Omega) = 1$, where Ω is "some physical state is present," is conventional. So the only issue is that of the additivity of propensity values, i.e., $\Pr(A \text{ or } B) = \Pr(A) + \Pr(B)$, when A, B are mutually exclusive states. Although I have no conclusive argument to show that this is universally true, it seems plausible for nonquantum systems. I doubt that countable additivity has any relevance for propensities.

References

Ballentine, Leslie. 2016. "Propensity, Probability, and Quantum Theory." *Foundations of Physics* 46: 973–1005.

Drouet, Isabelle. 2011. "Propensities and Conditional Probabilities." *International Journal of Approximate Reasoning* 52: 153–65.

Gillies, Donald. 2000. *Philosophical Theories of Probability.* London: Routledge.

Humphreys, Paul. 2004. "Some Considerations on Conditional Chances." *British Journal for the Philosophy of Science* 55: 667–80. Reprinted as chapter 10 of this volume.

———. 2007. "Andrei Nikolaevich Kolmogorov." In *New Dictionary of Scientific Biography*, edited by Noretta Koertge. New York: Charles Scribners and Sons.

———. 2018. "Knowledge Transfer across Scientific Disciplines." *Studies in History and Philosophy of Science.* Forthcoming. doi.org/10.1016/j.shpsa.2017.11.001

Lewis, David. 1980. "A Subjectivist's Guide to Objective Chance." In *Ifs: Conditionals, Belief, Decision, Chance and Time*, edited by W. L. Harper, R. Stalnaker, and G. Pearce, 267–97. Dordrecht: D. Reidel.

Lyon, A. 2014. "Rényi: There's No Escaping Humphreys' Paradox (When Generalized)." In *Chance and Temporal Asymmetry*, edited by Alastair Wilson, 112–25. Oxford: Oxford University Press.

---

[4] By admissibility I mean the sense given in Salmon 1967 that the propensities satisfy the standard axioms of probability theory, not the later sense of Lewis 1980.

[5] "Physical" here is taken in a very general sense to include anything that is spatiotemporally located. More general spaces could be used, although if they are, there seems to be no reason why we should disallow mental states as arguments in propensity functions.

McCurdy, C. S. I. 1996. "Humphreys's Paradox and the Interpretation of Inverse Conditional Propensities." *Synthese* 108: 105–25.

Miller, David. 2002. "Propensities May Satisfy Bayes's Theorem." In *Bayes's Theorem*, edited by Richard Swinburne, 111–16. Oxford: Oxford University Press.

Quine, W. V. O. 1951. "Two Dogmas of Empiricism." *Philosophical Review* 60: 20–43.

Reichenbach, Hans. 1949. *The Theory of Probability*. Berkeley: University of California Press.

Rényi, Alfred. 1970. *Probability Theory*. Amsterdam: North-Holland.

Salmon, Wesley C. 1967. *The Foundations of Scientific Inference*. Pittsburgh: University of Pittsburgh Press.

Suárez, M. 2013. "Propensities and Pragmatism." *Journal of Philosophy* 110 (2): 61–92.

———. 2014. "A Critique of Empiricist Propensity Theories." *European Journal for Philosophy of Science* 4 (2): 215–31.

Suárez, M. Forthcoming. "The Chances of Propensities." *British Journal for the Philosophy of Science*. https://doi.org/10.1093/bjps/axx010

Walley, Peter. 1991. *Statistical Reasoning with Imprecise Probabilities*. London: Chapman and Hall.

von Mises, Richard. 1939. *Probability, Statistics, and Truth*. London: Macmillan.

———. 1964. *Mathematical Theory of Probability and Statistics*. New York: Academic Press.

PART IV

General Philosophy of Science

Aleatory Explanations

What conditions of adequacy should we impose upon explanations of chance-like phenomena? A common response has been that, in addition to incorporating true laws, we should require that the explanation be probabilistic or statistical in form. "Probabilistic" and "statistical" have, unfortunately, been interpreted as entailing that somewhere in the explanation the correct probability value has to be cited, be it logico-inductive, frequentist, or strength of propensity.[1] This condition seriously overemphasizes the explanatory role of probabilities per se. Furthermore, it leads to imposing a completeness condition on probabilistically relevant factors which is unnecessarily strong. I shall show here that causal explanations of singular events, within which probability is used as an auxiliary device rather than an end, can be cumulative in a way that value-oriented explanations cannot, and in the course of showing this provide a simple solution to an intuitively puzzling aspect of aleatory explanations.

To avoid confusion, I must be quite explicit about the kind of causation I have in mind. It is the kind which satisfies the principle of relevant difference—that a cause must make some difference to the chance of an event's occurrence. Accounts of causality satisfying this principle have been given in Reichenbach 1956, Suppes 1970, Cartwright 1979, and Humphreys 1980. In the following discussion, I shall try to invoke only the basic principle that causes involve changes in probability and, to avoid bringing in subsidiary issues, I shall restrict my attention to direct causal antecedents of specific

---

[1] Theories of explanation which incorporate this view include Hempel 1965, 376–412, with an elaboration to be noted in Hempel 1968, 121–22; Salmon 1971, 76–77, 58–65; Fetzer 1974, 187, 191–93; and Railton 1976, 217–18, 221–22. The causal-relevance model outlined in Salmon 1978 retains this view (699). I should also note that Hempel discusses only what he calls the "logic of the simplest type of probabilistic explanation" (1968, 117) and that Jeffrey, although relying heavily on the use of correct stochastic processes in his 1971, states, "One may gloss this statement by pointing out that the actual outcome had such-and-such a probability, given the law of the process, but this gloss is not the heart of the explanation" (Jeffrey 1971, 24).

events. Roughly, X is a direct cause of Y if X is a cause of Y and there are no intervening causal factors between X and Y. (A precise definition of this notion can be found in Humphreys 1980, 309.) Other views about causality have been stated in connection with probabilistic explanation. Salmon's conception of causal explanation for example, involves causal propagation by a continuous process between cause and effect, with causal interactions playing a central role. Because the possibility of mark transmission is the distinguishing feature of causal processes in his approach, the principle of relevant difference is not always satisfied. The reader is urged to consult Salmon 1978, 1981 for details of that account. A central feature of that view, however, which is shared by Fetzer (1974) and Railton (1976) within their own causally oriented accounts of explanation, is that the true value of the probability incorporated in the relevant laws must be available for an adequate explanation.[2] It is this feature which I believe not only can, but must be, avoided.

Let me begin by recalling a feature of the statistical-relevance model[3] which many have found objectionable—the inclusion of negatively relevant factors in explanations. For example, as an explanation of why a car skidded off the road while rounding a bend, the assertion, "The car skidded off the road because it was traveling at an excessive rate of speed and the road was free of ice," would be acceptable within the statistical-relevance model, if the two factors cited exhausted the class of relevant factors present on that occasion. Yet specifying an ice-free road as an explanatory factor of an event whose occurrence is less likely to occur as a result of that condition seriously weakens the intuitive acceptability of the explanation and hence of the model to which it conforms.[4] The need to include such negative factors stemmed from the basic requirement that the explanandum event must be attributed the correct probability. Explanation of a particular event went by assigning it to the broadest homogeneous reference class describing it, and omitting any statistically relevant factors—be they negative or positive—would (usually) result in awarding an incorrect probability to that event. I have chosen the statistical-relevance model as an example because in its insistence that a solution to the problem of single-case probabilities is needed it exemplifies exactly the feature which these other explanatory accounts mentioned above also require—correct probability values.

The simplest way to avoid this problem would be to drop the true probability condition and either to require only positively relevant factors to be included in an explanation or to allow explanations only in cases where negative

---

[2] Others who have mentioned causal explanations are Cartwright (1979) and Skyrms (1980, 144). Their outlines appear to be compatible with much of what is said here.

[3] As given in Salmon 1971b.

[4] For typical objections to this inclusion of negative factors see Mellor 1976 and Cartwright 1979, 422–23, 425.

factors are absent.[5] The first tactic leads to misleading explanations, and the second is unduly restrictive.

Fortunately, there is an easy solution to the dilemma. First, recognizing the inadequacies of a purely statistical approach, let us insist that an explanation must cite causally relevant, rather than merely statistically relevant, factors. Then let us call positive or *contributing* causes those which raise the probability of an effect, and negative or *counteracting* causes those which lower the probability of an effect. (There are no neutral causes.) This immediately indicates that probabilistic causal explanations are likely to be fundamentally different in form from nonprobabilistic causal explanations, for there are no analogues of our counteracting causes in deterministic cases. When a traditional nineteenth-century determinist such as Mill wrote of "negative causes," for example, he meant the *absence* of those sufficient counteracting causes which if present would have—rather than might have—prevented the effect's occurrence.[6] The key feature of probabilistic counteracting causes is that they can co-occur with an event whose occurrence they tend to prevent, and this is not possible with sufficient counteracting causes.[7]

With this distinction between contributing and counteracting causes in mind, the difficulty encountered by the statistical-relevance model can be seen to stem from the uncritical extrapolation of the standard "because" format from the nonprobabilistic to the probabilistic case. Instead, the canonical explanatory form for probabilistic causal explanations is "*A* because φ, despite ψ," where φ is a nonempty set of contributing causes, ψ is a set, possibly empty, of counteracting causes, and *A* is a sentence describing what is to be explained. Thus the explanation in the example discussed above would be "The car skidded off the road because it was traveling at an excessive rate of speed, despite the fact that the road was free of ice." This is, I think, an acceptable explanation, yet complete in including all known factors.

Introducing a new linguistic form will not, of course, dissolve the philosophical arguments which have been leveled against probabilistic causality as a species of causation. Nevertheless, much of the concern about probabilistic explanation does, I think, stem from the conflict between the original models and ordinary use, and recognizing the correct linguistic form should go a long way toward avoiding that concern. As has been correctly pointed out, however, no satisfactory model of explanation should rely upon what we might

---

[5] Mellor (1976), for example, endorses the latter view.

[6] See Mill 1874, book 3, chap. 5, sec. 3.

[7] The co-occurrence is not always possible, even in probabilistic cases. The absence of a necessary condition for an effect will be a sufficient counteracting cause of that effect. Necessary conditions as a special case of probabilistic causality are discussed in Humphreys 1980, 310–11. I thank James Fetzer for bringing this particular case of a counteracting cause to my attention. Note that in this case Mill's "negative cause" would be a contributing cause for us.

call idiosyntactic forms of a particular language. Let me then make clear the structure of the explanatory sentence. "Despite" and its equivalents "although" and "even though" are often classified as conjunctions, but this cannot be the case here, for they are not commutative in this explanatory context. Hence an aleatory explanation is not simply a list of causes, positive followed by negative. In fact, the "despite" clause cannot itself be explanatory, for when φ is empty, there is no explanation, irrespective of how many elements ψ contains. The function ψ performs is, I believe, one of indicating in the simplest possible way how strong the explanation is. The greater the number of elements in ψ and the more negative they are, the weaker the explanation becomes for a fixed φ. To emphasize the fact that the negative factors are not themselves explanatory, I shall henceforth use the alternative form "*A* even though ψ, because φ," which has the merit of placing the counteracting factors outside the scope of the "because" clause. The preferred form is still, for ordinary purposes, the one involving "despite."

Thus far, nothing has been said which would exclude specifying the probability as well as the causes. The question of whether we have to arises when we consider if partial explanations are possible. Because almost all effects are the result of multiple causal influences, we must now ask whether φ and ψ have to be maximal (in the sense of including all contributing and counteracting causes) for an explanation to be acceptable. Here we must keep separate two related but distinct requirements traditionally imposed on explanations. First, the laws cited in the explanation should be true. From now on, I shall simply assume that this requirement is satisfied. Aside from the view found in all realist explanatory models that no genuine explanation can be based on falsehoods, we shall better see how the causal approach is superior to alternatives if we adhere to the condition of true laws. The second requirement has a number of variants, which I shall generically call "maximal specificity conditions," each of which insists that all relevant conditions which (are known to have) occurred on the occasion in question be cited in the explanation. The motivation for imposing maximal specificity is, I think, closely connected with the emphasis on correct probability values noted earlier. Having true laws, we need to know which is the appropriate one for the occasion in question, and a maximal specificity condition helps us do this. In particular, if we fail to specify relevant factors which were present on the occasion in question, our explanation is open to being overthrown by later inclusion of those factors, because the original law, although true, was not true of the situation under consideration. Given that we are rarely cognizant of all the causal influences present on a given occasion, does omitting to mention some of them through ignorance forever make causal explanations potentially defeasible? I shall argue that causal explanations are in a stronger position here than explanations requiring probability values, for while the omission of even a single relevant factor in the latter models will (usually) render these explanations incorrect, causal models can fail (through epistemic limitations) to specify many relevant factors, yet still provide perfectly acceptable, albeit improvable, explanations.

To bring out the contrast with approaches which require probability values, first note that within the theory of probabilistic causality used here (where the main emphasis is on increases and decreases in probability), all causal claims could be made using only comparative probabilities, where only order relations are used. Let us suppose, however, that we do attach numerical values to the conditional probabilities. Unless X is the only factor affecting A's probability, when we compare $P(A|X)$ with $P(A|\sim X)$, the probability functions $P(\cdot|X)$ and $P(\cdot|\sim X)$ will be averaged over some set of probabilistically relevant factors R. In many cases, although not all, this will not matter. Suppose we discover that on the occasion in question, A was preceded by Y as well as by X, and that $P(A|XY) \neq P(A|X)$. It may well still be true that $P(A|XY) > P(A|\sim XY)$ whether Y is itself a counteracting or contributing cause of A, and hence we can still claim that in the circumstances X was a contributing cause of A. To return to our example of the skidding car, the initial explanation was: "The car skidded off the road even though the road was free of ice, because it was traveling at an excessive rate of speed." Discovering later that the car's brakes were defective but that it was broad daylight when the accident occurred, we have an expanded explanation: "The car skidded off the road even though visibility was good and the road was free of ice, because the car was traveling at an excessive speed and had defective brakes." Here our augmented explanation does not destroy the original explanation, nor show it false, because within the circumstances, defective brakes and excessive speed were contributing causes, while good visibility and an ice-free road were counteracting causes. This holds even though it is highly likely that the probability value of a skid given all four factors differs from the value given only the two original factors. Even were we to discover a multitude of counteracting causes which lowered the probability of a skid to some small value, those original factors would still be explanatory.

The situation is not, of course, always so straightforward. Not all causes are invariant in the manner of those mentioned above. Single doses of a fertilizer may increase crop yield, but double or successive doses may easily diminish it. Also, once we leave direct causes, intervening factors may destroy causal chains in the way discussed in Humphreys 1980. Such interactive effects[8] may well result in a transfer of factors from φ to ψ, or vice versa, or eliminate them entirely, perhaps leaving φ empty and thus leaving us with no explanation at all. How are we to avoid this possibility? It looks as though we require maximal specificity again and will ever be open to defeasible explanations. The situation

---

[8] I emphasize here that the difficulty is one of actual causal interaction and not one of deficient statistical methodology.

is not so extreme, however. Recalling that (as with other models) we insist on true laws, we need what Bromberger[9] has called "general abnormic laws," laws here of the form "X's raise the probability of A's unless Γ," where Γ contains a specification of all those factors which convert X's into counteracting causes or else neutralize them. What needs to be known about the specific circumstances, therefore, is whether any elements of Γ were present. We can be in ignorance of, and hence fail to specify in an explanation, any number of factors which raise or lower the probability, just so long as they do not defeat X as a contributing cause—or, analogously, Y as a counteracting cause—of A in the circumstances.

As a practical matter, I believe that much of our causal and explanatory knowledge is of this form—that we know, within the limits of certitude attending *any* law, that X's cause A's when certain common defeating conditions are absent, and, having checked that such conditions are absent, we have a partial explanation for A upon which we may later improve. Additional research may locate other factors present on that occasion, but that is reason for adding to, rather than destroying, the original explanation. Of course, everything argued here depends upon the number of defeating conditions being finite, but if that were not the case, we could not even formulate the law. For pragmatic reasons, some kind of maximal specificity condition still ought to be imposed upon an explanation, for omitting known factors may well give a misleading impression of its completeness. The important difference lies in the security we have against the effects of hitherto unknown factors.

What do we lose by forgoing probability values? Only, perhaps, a sense of expectation induced by the values. But it has been argued sufficiently often by now that predictive power is not something that should be expected of an explanation that this loss should not concern us. To those arguments I would add one further. If we examine Grünbaum's admirable defense of the "symmetry thesis" (Grünbaum 1973), it becomes clear that this thesis is simply based on the truism that logical relations—be they deductive or inductive—are temporally invariant. When it emerged that inductive explanations were essentially epistemically relativized, the claim that epistemological asymmetries had nothing to do with this thesis should have become immediately suspect. The logical relation remains invariant, it is true, but the possibility of having the knowledge set used in the explanation at the time a prediction is needed becomes a truly counterfactual possibility in most probabilistic cases. We need, therefore, not be concerned by the lack of a probability value.

---

[9] The full formulation of abnormic laws is given in Bromberger 1966. The widespread use of such laws indicates, I think, that the conditions imposed by Cartwright on causal laws in 1979, 423, are too strong.

References

Bromberger, Sylvain. 1966. "Why-Questions." In *Mind and Cosmos: Essays in Contemporary Science and Philosophy*, edited by Robert G. Colodny, 86–111. Pittsburgh: University of Pittsburgh Press.

Cartwright, N. 1979. "Causal Laws and Effective Strategies." *Noûs* 13: 419–37.

Fetzer, James H. 1974. "A Single-Case Propensity Theory of Explanation." *Synthese* 28: 171–98.

Grünbaum, Adolf. 1973. "The Asymmetry of Retrodictability and Predictability, the Compossibility of Explanation of the Past and Prediction of the Future, and Mechanism vs. Teleology." In *Philosophical Problems of Space and Time*, 2nd ed., edited by Robert S. Cohen and Marx W. Wartofsky, 281–313. Dordrecht: D. Reidel.

Hempel, Carl. 1965. *Aspects of Scientific Explanation and Other Essays in the Philosophy of Science*. New York: Free Press.

———. 1968. "Maximal Specificity and Lawlikeness in Probabilistic Explanation." *Philosophy of Science* 35: 116–33.

Humphreys, Paul. 1980. "Cutting the Causal Chain." *Pacific Philosophical Quarterly* 61: 305–14.

Jeffrey, R. C. 1971. "Statistical Explanation vs. Statistical Inference." In Wesley C. Salmon, *Statistical Explanation and Statistical Relevance*, with contributions by Richard C. Jeffrey and James G. Greeno, 19–28. Pittsburgh: University of Pittsburgh Press. Originally published in *Essays in Honor of Carl G. Hempel*, edited by Nicholas Rescher et al. (Dordrecht: D. Reidel, 1969).

Mellor, D. H. 1976. "Probable Explanation." *Australasian Journal of Philosophy* 54: 231–41.

Mill, John Stuart. 1874, *A System of Logic*. 8th ed. New York: Harper and Bros.

Railton, Peter. 1976. "A Deductive-Nomological Theory of Probabilistic Explanation." *Philosophy of Science* 45: 206–26.

Reichenbach, Hans. 1956. *The Direction of Time*. Berkeley: University of California Press.

Salmon, Wesley C. 1971. "Statistical Explanation." In *Statistical Explanation and Statistical Relevance*, with contributions by Richard C. Jeffrey and James G. Greeno, 29–87. Pittsburgh: University of Pittsburgh Press. Originally published in *Nature and Function of Scientific Theories*, edited by Robert G. Colodny (Pittsburgh: University of Pittsburgh Press, 1970), 173–231.

———. 1978. "Why Ask 'Why'?" *Proceedings and Addresses of the American Philosophical Association* 51: 683–705. Reprinted in *Hans Reichenbach: Logical Empiricist*, edited by Wesley C. Salmon (Dordrecht: D. Reidel, 1979), 403–25.

———. 1981. "Causality: Production and Propagation." In *PSA: Proceedings of the Biennial Meeting of the Philosophy of Science Association 1980*, edited by Peter D. Asquith and Ronald Giere, vol. 2, 154–71. East Lansing, MI: Philosophy of Science Association.

Skyrms, Brian. 1980. *Causal Necessity*, New Haven: Yale University Press.

Suppes, Patrick. 1970. *A Probabilistic Theory of Causality*. New York: North-Holland.

{ 13 }

Analytic versus Synthetic Understanding

On many accounts of explanation, the purpose of scientific explanations, indeed of explanations of any kind, is to produce understanding. Hempel was skeptical of psychologically oriented accounts of understanding in his early writings on explanation;[1] scientific understanding, on the other hand—which revealingly he identified with theoretical understanding—was the goal of explanatory activity:

> Our main concern has been to examine the ways in which science answers why-questions of the [explanatory] type and to characterize the kind of understanding it thereby affords. . . . The understanding it conveys lies . . . in the insight that the explanandum fits into, or can be subsumed under, a system of uniformities represented by empirical laws or theoretical principles. . . . The central theme of this essay has been, briefly, that all scientific explanation . . . seeks to provide a systematic understanding of empirical phenomena by showing that they fit into a nomic nexus. (Hempel 1965a, 488)

Perhaps because nontheoretical modes of understanding seemed to Hempel to be incapable of precise treatment, perhaps because he was skeptical of "empathic understanding" and the continental tradition of *Verstehen*, Hempel, and almost all of those who wrote after him, approached the concept of scientific understanding indirectly through an analysis of the concept of scientific explanation.[2]

It is useful to see what results when we address scientific understanding rather more directly, using recent work in explanation as an occasional

---

[1] See, e.g., Hempel 1942, sec. 6; 1948, sect. 4. All page references to Hempel's writings cited here refer to Hempel 1965b.

[2] A number of philosophers have attempted to capture the process of how scientific explanations provide understanding. For example, Friedman 1974, Achinstein 1983, Salmon 1984, and Kitcher 1989 all recognize the fundamental importance of understanding in the scientific enterprise.

resource. We shall see that scientific understanding provides a far richer terrain than does scientific explanation and that the latter is best viewed as a vehicle to understanding, rather than as an end in itself. My purpose here is primarily to explore what I call "analytic modes of understanding," although these modes sit uneasily with Hempel's own work, especially with his deductive-nomological model.

13.1. Analytic and Synthetic Understanding

In the Port-Royal Logic, Antoine Arnauld and Pierre Nicole wrote: "there are two kinds of method, one for discovering the truth, which is known as *analysis*, or the *method of resolution*, and which can also be called the *method of discovery*. The other is for making the truth understood by others once it is found. This is known as *synthesis*, or the *method of composition*, and can also be called the *method of instruction*" (Arnauld and Nicole [1683] 1996, 233).[3]

In a similar vein, the seventeenth-century Cartesian Pierre-Sylvain Régis claimed that there were two methods "of which one serves to instruct ourselves and is called analysis . . . and the other which is used to instruct others is called *synthesis*" (Régis 1690).[4]

The central distinction brought out by these writers is one of great importance.[5] Understanding comes initially through discoveries, and the techniques that science has developed for discovery are not always the same as those it uses for conveying existing knowledge to subsequent inquirers. At the time that Hempel formulated his views on explanation, justification was a legitimate topic of inquiry whereas discovery was less respectable. In both the deductive-nomological and the inductive-statistical models of explanation there was a parallel, most dramatically exhibited in the symmetry thesis, between the justificatory uses of nomological arguments and their explanatory uses. Because of this parallel, which I shall examine in greater detail later, it was no accident that Hempel's model of explanation took on a synthetic rather than an analytic form. For synthetic arguments and justificatory arguments have the same form in Hempel's approach, and indeed Hempel frequently views explanations

---

[3] For Arnauld and Nicole the synthetic method is the more important of the two "since it is the one used to explain all the sciences" [(1683) 1996, 239].

[4] In the case of mathematics, the view that analysis was personal may have arisen from the puzzlement, dating back to Pappus's claims about analysis, concerning how a "backward" procedure could yield certain knowledge, or could be a legitimate route to knowledge at all.

[5] Arnauld and Nicole identify three things: a method of analysis, a process of instructing oneself, and a method of discovery. The Port-Royal was certainly on the right track, but it is better to keep these three things separate, for self-instruction is not limited to analysis, nor to discovery; analysis can enlighten others, as can a discovery; and analysis does not always lead to discovery, whereas synthesis itself can be a discovery device.

TABLE 13.1 Some principal approaches
to explanation

Analytic	Synthetic
Causal	Covering law
Functional	Unification

as objects designed to convey understanding to individuals other than the one providing the answer to the explanatory why-question. (I do not mean that these were the primary reasons for Hempel's adopting a synthetic approach, but simply that by so doing, the resulting apparatus fitted easily into the prevailing views about discovery versus justification.)

What should we mean by an "analytic process"? An analytic process resolves a complex entity into a multiplicity of constituents. Those constituents may or may not be ultimate, where ultimate constituents are those elements that cannot be further resolved by the methods used in the analysis: ideally an analytic process does reach such ultimate components. What is ultimate is relative to the methods used within a field of inquiry. For example, the possibility of the conceptual division of an elementary particle of physics does not undermine that particle's claim to be ontologically ultimate, nor does the spatial division of a syntactically fundamental object destroy that object's primitive syntactic nature. The constituent elements arrived at by analysis must also be independent. This independence may be logical, causal, statistical, functional, or some other kind appropriate to the elements involved. There are thus different kinds of analysis: definitional analysis, causal analysis, statistical analysis, functional analysis, and so on. A synthetic method, in contrast, is one that produces a given entity by means of a combination of elements or constituents.[6] With this rough characterization, we can classify the main contemporary approaches to explanation (table 13.1).[7]

---

[6] Not all synthetic objects, that is objects produced by a synthetic method, can be analyzed, nor can all analytically decomposed objects be (re)synthesized.

[7] Examples of the causal approach can be found in Salmon 1984 and Humphreys 1989; of the covering law approach in Hempel 1965a; of the functional approach in Cummins 1977; and of the unification approach in Friedman 1974 and Kitcher 1989. I set aside here pragmatic accounts of explanation. In laying out this taxonomy I am in no way denying the validity of the tripartite division used by Wesley Salmon in his 1984 and 1989 works, that division being between the ontic, the epistemic, and the modal conceptions of explanation. That is an extremely important taxonomy and one that is consistent with the one used here. Salmon's classification does, however, divide the conceptual space in a rather different way than does my own.

13.2. Primary and Secondary Understanding

Many have followed Hempel (and Bromberger) in construing explanation as consisting in an adequate answer to a why-question.[8] Judging from the way that contemporary philosophers write about erotetic issues, it is standard to construe the explanatory dialogue as involving one individual posing the why-question and a different individual providing the answer. Moreover, probably because the need to provide plausible examples requires it, there is a tendency to think of these issues in terms of the answer to the why-question already existing, usually somewhere in the scientific literature.[9] These assumptions are entirely reasonable in most cases, because ordinarily the questioner suffers from an epistemic deficit that the respondent does not, and because almost all of the why-question approaches are linguistically oriented, one instinctively thinks of the answers as being "already on the books."[10] Yet there is nothing in the nature of why-questions themselves that requires the questioner and answerer to be different individuals. Furthermore, the most important kinds of why-questions are those asked at a time when no answer to them is known. Let us then call *primary understanding* the (increase in) understanding that is achieved when an individual or a group discovers at least one previously unknown entity that allows progress toward an (improved) explanation of some phenomenon or, without aid from others, an individual or a group rediscovers the steps made by an earlier discoverer. (This does not include simply looking up an explanation that has been written down.) Primary understanding thus arises at least in part from a method of discovery. We can call *secondary understanding* the (increase in) understanding that occurs when one individual or group is provided by another individual or group with all the items of knowledge that the first needs in order to arrive at some (improved) level of understanding. (A common source of secondary understanding is the scientific literature, where the role of the provider is indirect.) Secondary understanding thus involves a method of instruction.

The distinction between the two types of understanding applies to a wide variety of cases. Primary understanding can be achieved by the first person to carry out a previously undiscovered Hempelian deduction from known laws and facts, by the discovery of a hitherto unknown cause, or by the discovery of a new set of unifying principles in a science (as for example Kepler did by discovering the laws that were later named after him). It is not necessary

---

[8] One can do this even when, as I do, the why-question is not the starting point of the explanatory process but is merely a common companion of the process.

[9] Why-questions can be viewed as abstract entities. Nevertheless, their role is usually that of one part of a dialogue between epistemically imbalanced agents.

[10] Kitcher's 1989 use of the phrase "the explanatory store" is a reflection of this.

for the provider of primary understanding to have first asked the associated why-question.

Primary understanding can be cooperative in form but its orientation is inward, not outward, in that the understanding desired is initially for the investigator(s) and not for others. Primary understanding is the basis of all secondary understanding in the sense that the latter could not occur without the former but the secondary understanding that builds upon the initial discovery does not always have the same form as the primary understanding that makes it possible. Frequently "textbook" explanations are different from, clearer than, and better than, the explanations given by the pioneers. It is an important question to ask, but one whose answer I shall not explore here, about how primary and secondary understanding are related.

We have already seen from the short list of possibilities that I cited in the last paragraph that primary understanding need not involve analysis, and secondary understanding need not involve synthesis, as the authors of the Port-Royal suggested. From now on, however, I shall restrict my attention for the most part to analytic primary understanding.

Hempel's own deductive-nomological account is indifferent between primary and secondary understanding; it would count as a deductive-nomological explanation the first time that Newton applied his laws of celestial mechanics to explain the motion of the moon as it did in subsequent repetitions to eighteenth-century why-questions. The important thing for our purposes is that primary understanding involves a reorientation from the Hempelian emphasis on justifying arguments to a process of discovery. Hempel's emphasis on justification is closely allied with his commitment to synthetic modes of understanding, and to understand how that commitment came about requires a brief examination of Hempel's treatment of why-questions.

13.3. Explanation as Discovery

Within Hempel's approach, as we all know, explanation, and hence understanding, were based on the availability of *nomic expectability*, a concept that led to most of the famous features of the covering law approach—the claim that explanations were arguments, the notorious high-probability requirement for statistical explanations, and the thesis of the symmetry of predictions and explanations, to name only three. Much has been made of this symmetry between explanation and predictions, but I want to draw your attention to a distinction that was more fundamental for Hempel, and which has been overshadowed by the symmetry thesis.

On the second page of his seminal essay "Aspects of Scientific Explanation," Hempel drew the distinction between explanation-seeking why-questions and epistemic why-questions. The former can be put in the standard form "Why is it the case that p?," where p is some empirical statement specifying the explanandum, whereas epistemic or reason-seeking why-questions have the form "What reasons are there for believing p?" Having contrasted these two types of why-questions, Hempel promised to establish that an answer to one should serve as an answer to the other: "as will be argued later . . . any adequate answer to an explanation-seeking why-question . . . must also provide a potential answer to the corresponding epistemic question" (Hempel 1965a, 335). But when one turns to the sections in which this argument is supposed to be given (secs. 2.4, 3.5), what one finds is an extended discussion of the relations between explanations and predictions instead of a discussion of the relations between answers to the two kinds of why-questions. There is only one sentence[11] about why-questions in the course of eighteen pages of discussion of the symmetry thesis. This leads me to believe that Hempel thought that all Hempelian answers to explanatory why-questions were explanations and vice versa, and that all Hempelian answers to epistemic why-questions were predictions and vice versa. The first of these equivalences is, as we now know, false. The second equivalence is our concern here.

Hempel opens and closes his essay with the contrast between the two types of why-questions, and it seems that for him this was the more important distinction, despite the fact that the explanation/prediction symmetry thesis received so much more attention from him and from his commentators. Yet by requiring explanations to have the same logical form as predictions, explanations were forced to take on a synthetic form that is too narrow to capture all legitimate kinds of scientific understanding. I shall now show that we should, as Hempel intended, take this contrast between the two types of why-questions as the more fundamental distinction, and that we should focus on the differences in the *epistemic* state of the two kinds of inquirers, rather than on the supposed sameness of *logical* structure of the vehicles that convey the requested information.

Let us restrict ourselves to deductive-nomological explanations of singular events for the moment. Hempel construed the differences between explanation and prediction in such cases as due only to what he called "certain pragmatic respects" (Hempel 1965a, 366–67; see also Hempel 1948, 249).

> In one case [i.e. explanations], the event described in the conclusion is known to have occurred and suitable statements of general law and particular fact are sought to account for it; in the other [i.e. predictions], the latter statements are given and the statement about the event in question is derived from them. (Hempel 1965a, 367)

---

[11] Hempel 1965a, 368.

But it is very strange to give, as Hempel did, a pragmatic interpretation to these differences because they are clearly epistemic in nature and not pragmatic. Although, for Hempel, the logical form of an explanation is identical to the logical form of a prediction, the epistemic state of an epistemic why-questioner clearly differs from the epistemic state of an explanatory why-questioner. To see this, consider the form of an Hempelian explanation.

Any Hempelian explanation or prediction of a particular fact involves a quadruple, the structure of which is <laws, particular facts, a derivation, a conclusion>. As Hempel's quote cited above indicates, an explanation seeker knows the truth of the last of these, but is ignorant of at least one of the first three. A prediction seeker, according to Hempel (see above quote), knows the first two and is searching for the third, which will, in the deductive contexts with which we are here concerned, give him the fourth. The situation is quite different with regard to the contrast between the epistemic state of the two kinds of why-questioners. In contrast to prediction seekers (as conceived of by Hempel), those seeking an answer to an epistemic why-question of the form "What reasons are there for believing that p?" will, as well as not yet believing the conclusion, be ignorant of at least one of the first three elements of the quadruple, and often, of at least one of the first two.[12] So the epistemic state of epistemic why-questioners will often be different from the epistemic state of Hempelian prediction seekers and, consequently, how the knowledge state of an epistemic why-questioner is altered by a satisfactory answer to his or her question will frequently be different from the way in which the knowledge state of a Hempelian prediction seeker will be changed. That is, the *new* knowledge acquired by an epistemic why-questioner will frequently be different from the new knowledge acquired by a Hempelian prediction seeker. But more importantly, the one thing which *always* differentiates explanatory why-questioners from epistemic why-questioners is their epistemic state with respect to the conclusion, which the former always knows but the latter does not know and may not even believe.

Of course, we could just abandon Hempel's restrictive account of prediction seeking—the case where the only thing that is missing is the deductive argument—and allow that a prediction seeker could be in the same kind of broader epistemic deficit as is an epistemic why-questioner, where any of the first three members of the quadruple might be missing. But this then makes it quite clear that what really separates the two kinds of inquirer, the justification seeker from the explanation seeker, is the difference in their status vis-à-vis the last element of the quadruple.

---

[12] I do not assume that the set of known sentences is closed under deductive consequence. If it is, and Moore's Paradox (I know x but do not believe x) is ill-founded then nonbelief in the conclusion entails the epistemic lack of at least one of the first three.

This epistemic asymmetry has consequences, for very often information contained in the explanandum itself provides vital clues for obtaining a primary understanding of that explanandum and these clues can be discovered by analysis of something already in the possession of the explanation seeker. Yet by virtue of providing a counterfactual characterization of the similarities between epistemic and explanatory why-questions ("the explanatory argument might have been used for a deductive prediction of the explanandum event *if* the laws and particular facts adduced in the explanans had been known and taken into account" (Hempel 1965a, 366), this essential epistemic asymmetry was erased, and explanations were forced into the synthetic mode.

So let us take the key feature of explanation seekers to be that they know the truth of the explanandum, that this sometimes means that they have access to the explanandum phenomenon itself, and they are using that as the starting point of their search for understanding.

13.4. Analytic versus Synthetic Understanding

Existing accounts of explanation divide into two basic kinds, and our examination of the epistemic differences between answers to the two types of why-questions, explanations, and predictions suggest what they are. In the Hempelian prediction case, the epistemic direction follows a process from the given facts, which in the D-N case are already sufficient for the explanandum sentence's truth, to that conclusion, and the task is to construct an argument that will direct that epistemic process. This is no trivial task of course, but it is essentially synthetic in form, by virtue of its combining already given components to produce a dependent explanandum sentence. But in the case of the explanation-seeking why-questions, the process is initially "backward" and consists in a process of discovery, seeking which laws and/or specific facts would give rise to the given explanandum. The discovery process is relatively trivial in the case of secondary understanding (you find an expert, or a reliable source of information), but it is crucial in the case of primary understanding. It is here that analytic methods tend to play an important role, although as we have seen they are certainly not the only method of attaining primary understanding.

We can most clearly see how such analytic understanding takes place by looking at some simple cases of causal analysis.

13.5. Explanatory Refinement and "How Exactly" Questions

Within the context of causal explanations it is natural to think of the explanandum event as an effect of some cause, but it is important that we not identify explananda with effects in causal contexts. In the canonical form of an explanatory why-question, "Why is it the case that p?," the sentence "p" can have the form "X is the cause of Y" or "X was involved in bringing about Y."[13] Thus, in cases where we have the explanandum itself available, the question could be about a cause, about a process connecting a cause with its effect, or, as we shall see, about entities containing one of these two things.

Understanding is not an all-or-nothing affair and an adequate account of scientific understanding must describe ways in which we can increase that understanding. Analysis then plays an important role in *explanatory refinement*. A common variety of explanatory refinement takes some part of the world that we suspect plays a role in producing a particular phenomenon and identifies the precise etiological agent that caused the outcome. This variety comes in two forms. The first takes a composite object and identifies which substance(s) within it are the "active ingredients." Here the analysis separates a compound object into its parts. For example, we say that smoking causes lung cancer. Yet there are known to be more than 3,800 components of cigarette smoke, and chemical analysis is a first and important step toward identifying the carcinogens among those thousands of components. A historically important example of this use of causal analysis was the discovery by William Withering in 1775 of digitalis as a treatment for dropsy (edema resulting from progressive heart failure). Withering took his cue from a folk remedy administered by an old woman in Shropshire, and from the twenty or so ingredients in her secret recipe, he isolated the foxglove herb as the effective component. (Withering's analysis has since been further refined to identify digitalis glycoside as the active ingredient in foxglove leaves.)

The second kind of analysis identifies which of the many properties possessed by an object are the ones responsible for bringing about an effect. To take a simple example, if a crumbling ledge on a New York City apartment building collapses and a clay flowerpot and a stone gargoyle of identical masses injure separate pedestrians on the street below, it is the kinetic energy transferred from the falling objects to the pedestrians that injures them and not the objects qua flowerpot or qua gargoyle, nor most of the other properties possessed by those objects.

13.6. Understanding Does Not Require a Mechanism

A number of authors have suggested that scientific explanation consists, at least in part, in displaying the mechanisms that lead up to the explanandum

---

[13] In cases where this is false, and X is actually not involved in bringing about Y, then the presupposition of the why-question is false and suitable corrective answers are in order.

event. I originally resisted adopting this view because it was clear to me that there were serious difficulties in applying it to quantum systems and, in a quite different way, to economic and social systems. I now believe that there is a persuasive set of more general reasons why being able to display a mechanism should not be a requirement for either understanding or explanations. Consider the case of aspirin. The Assyrians knew that chewing willow leaves produced analgesic effects, and Hippocrates used extract of willow bark as a pain reliever. The active ingredient of these folk remedies, salicin, was first isolated in 1828–29 by Buchner and Leroux, with a further refinement to salicylic acid in 1838 by Piria. The discovery and isolation of this material contributed to Felix Hoffman's synthesis of acetylsalicylic acid in 1897, a discovery that led to the first commercial manufacture of aspirin by the Bayer company. For seventy years there was an overwhelming amount of evidence from both scientific trials and everyday usage that acetylsalicylic acid caused relief from headaches and from fever in a wide variety of populations and biological contexts. One thus had exactly the kind of invariant connection between a factor and an outcome that is characteristic of genuine causal relations. Thus it would have been entirely correct to have claimed in say, 1920, that one understood why a particular individual's headache had disappeared, that understanding consisting in the knowledge that the individual had taken some aspirin, and that taking aspirin almost invariably leads to cessation of headaches, even though the mechanism of action between the cause and the effect was unknown at that time. For it was not until the work of the British pharmacologist John Vane in the 1970s that the mechanism of action of aspirin (inhibition of the synthesis of prostaglandins) was discovered.[14]

We have an explanation behind the understanding, too, because a cause of the headache's cessation has been identified, even though the operative mechanism has not. In cases where we have substantial evidence for the invariance of the association between cause and effect, as we do with aspirin, the need for knowledge of a mechanism can be dispensed with.

To take a different example, there is an obvious process connecting smoking and the site of lung cancer, yet the mechanism of carcinogenesis produced by that smoke pathway is still not properly understood at the present time, nor is the exact way in which sexual intercourse is linked to an increased risk of cervical cancer in women.[15] Such examples are by no means limited to the life sciences. The initial lack of a known mechanism for gravitation, for heat, and for electricity did not preclude early investigators from acquiring a considerable degree of understanding of the relevant phenomena, and especially of their causal effects.

---

[14] This mechanism is at present not fully understood.
[15] I owe this last example to David Freedman.

It is certainly true that knowledge of such mechanisms tends to further our understanding in ways that would be difficult without them. For example, Vane's theoretical work not only allowed separation of aspirin's effects into the analgesic (pain-killing), the anti-inflammatory, and the antipyretic (fever reduction), as well as a certain amount of unificatory understanding of similar analgesics such as ibuprofen through similar modes of action, but it led to the realization that aspirin was also effective against recurrence of strokes by virtue of preventing buildup of prostaglandins at the site of arterial lesions. Equally importantly, mechanisms are the most important way of ruling out possible confounders (factors that are associated with the effect but are not causally responsible for it.)

13.7. Carriers and Causes

In many cases of causal discovery, what I shall call the *carrier* of the cause is identified before the cause itself is isolated.[16] Here is one famous example. Between 1849 and 1855 the London physician John Snow identified water contaminated by fecal matter as the carrier for cholera.[17] He had, as he put it, discovered the "mode of transmission" of the disease, but Snow never identified the specific microorganism that was responsible for cholera. That discovery was not made until 1883 when Robert Koch identified the *Vibrio cholerae* bacterium as the cause. This kind of situation is different from the one we discussed earlier involving aspirin. There the active causal agent (the acetylsalicylic acid) was known early in the scientific investigation in addition to the carrier. In the cholera cases, Snow had knowledge only of the carrier and not of the causal agent.

In the second kind of situation, we may ask this question: Do we have an explanation of the disease's occurrence when we have identified only the carrier? There is also a different question: Do we have some understanding of how the disease comes about when we have found the carrier? For those of us who adhere to a causal account of explanation, the answer to the first question has to be "No." Except in an elliptical manner of speaking, rats do not cause bubonic plague and neither do fleas, and polluted water is not the cause of cholera. It is the *Yersina pestis* organism that is the "active ingredient" of plague transmission and the rat is no more a cause of an illness than the piece of paper upon which this sentence is printed is a cause of your thinking rather differently

---

[16] When the carrier is a biological agent, it is known as a vector.

[17] Snow's treatise *On the Mode of Communication of Cholera* (Snow [1855] 1936) is a beautiful example of systematic scientific detective work, and the basis of one of the most important public health advances ever made. The full flavor (literally) of the often unspeakably filthy conditions in which most people lived prior to modern public sanitation is vividly conveyed in Snow's essay.

about explanation from now on. It cannot be, for the sentence would have had exactly the same effect had you viewed it, as I did, on a computer terminal. A factor, the presence or absence of which makes no difference to an effect, cannot be a cause of that effect.

Yet we do gain some measure of understanding through learning of these vehicles, for the discovery of the carrier has brought us closer to discovering what the explanation of the effect is, even though an explanation is not yet available. *We can thus have an increase in understanding without (yet) having an explanation.* We understand how the disease is transmitted but not what causes it (and thus, at least on the causal accounts of explanation, we have not explained why the individual had it).

It might be argued that polluted water is in fact a cause of cholera.[18] To address this concern, let us clarify the first question that we asked earlier. If the question is one involving a generalization—what, in general, causes cholera—then the answer to the above question is still "No." For polluted water is not a general explanation for cholera. There are other carriers including contaminated shellfish and excreta from infected individuals. What is invariant across all these carriers is the vibrio bacterium and it is that invariance which underpins the causal explanation resulting from Koch's discovery. However, if the question is about whether the carrier is part of the explanation in this particular case, the issue is more complicated, for what constitutes a correct answer to the question requires answers to a collection of other questions involving both factual and conceptual matters.

To argue that in specific cases the carrier plus the agent is a cause of the effect is to resist the persuasive line of argument that supports relevance as a key feature of causation, and hence of causal explanations. The failure of explanatory relevance within Hempel's deductive apparatus was highlighted by Henry Kyburg's hexed salt example.[19] The argument "All samples of hexed salt placed in warm water dissolve. This sample of salt was hexed and placed in warm water. Therefore, this sample of salt dissolved." satisfied all of Hempel's criteria of adequacy for D-N explanations, yet it is clearly not an explanation. In order to produce an effective example, Kyburg had to explicitly display the irrelevant predicate "hexed" separately from the relevant predicate so that the syntactic structure of the premises already represented an analysis of factors into those that were relevant and those that were not. It is, needless to say, a much more common situation in science for a factor cited in an explanation to be identified by means of a predicate that does not represent such a separation.

One reason for the difference in views about contaminated drinking water being a cause of cholera doubtless lies in a different choice of contrast

---

[18] I have heard this reaction from a number of different audiences.
[19] See also the important examples given in Salmon 1971.

cases for the causal analysis. If one contrasts the actual situation containing contaminated water with a situation within which the water plus bacteria is absent, then one arrives at the position that the polluted water is indeed a cause of the disease. If, alternatively, one uses as a contrast the case where the water is still present but the contaminating bacteria are removed, then the water cannot be a cause. Which contrast case is the correct one? Perhaps minimal change semantics for counterfactuals could decide, but a more objective criterion employs experimental analysis on the polluted water itself. Analyze the medium that was present in the case concerned into the carrier and its various pollutants, then see whether the disease occurs in the presence of each single component. Obviously, an inductive inference is required to move from the experimental case to the original case, but with a well-designed experiment this is a safer move than one relying on similarity metrics between possible worlds. (Snow's own use of a naturally randomized experiment, by the way, was incapable of isolating the bacterium, a fact that casts doubt on the ability of many epidemiological trials to identify the cause rather than the carrier.)

I said that Koch discovered the cause of cholera. This is misleading, for current evidence suggests that it is a toxin manufactured by the *Vibrio cholerae* bacterium that is actually the cause of the disease. Such cases raise an obvious question: Are the only causes those factors for which no further decomposition into relevant and irrelevant components is possible? My answer is, tentatively, "Yes." If one is willing to put aside what is ordinarily said, as one should, then the argument from analysis just given surely leads us to that conclusion.[20]

The central issue raised in this section is, I think, one of quite general applicability. Consider just one example pointed out to me by Eric Scerri. A commonly used mechanism for explaining how our olfactory senses detect various smells is the "lock and key" account of molecular action. Only a small part of the molecule is involved in the detection; is the entire molecule or simply the segment "actively involved" in the process of smelling a cause of our smelling perfume? More prosaically, with an ordinary key is it the entire key or just the cut-outs (or magnetic strip) that explains why the door opened? Our analysis above suggests that the key handle or the card is simply a carrier to convey the cause to where it can do its work.[21]

---

[20] I say "tentatively" here because the decision relies on the acceptance of a considerable number of assumptions about causation, the truth of some of which is not as certain to me as I should like.

[21] In other cases, such as the role of "junk DNA," I am not in a position to provide an answer.

13.8. Linguistic versus Material Analysis

A major drawback to any linguistically oriented approach to explanation is thus that it is not effective for the kind of discoveries that material analyses can lead to. Material analysis of objects can reveal hitherto unknown components for which the language as yet has no terms. The analytic decomposition of syntax is a poor cousin of material analysis, for language frequently does not reflect material structure, whereas the system itself trivially does.[22]

It bears noting that the subdivision of reference classes also has a serious defect, for that method never allows us to "go down" to lower-level properties or entities.[23] If we begin with samples of water as the members of the reference class, we shall end with samples of water as members of the reference class, even if the "relevant" subclass contains samples of water with highly specific characteristics. If the water is simply a carrier, then this method will partially misidentify the explanatory agent.[24]

We can see here one of the major differences between the linguistic approaches to explanation and material approaches. With the examples we have discussed of foxgloves, tobacco smoke, contaminated water, rats, and so forth, the object of potential understanding—the explanandum itself, not the explanandum sentence—is available for analysis whatever its mode of presentation, and this availability is crucial for the discovery process. It is only when one is trapped in a mode of secondary linguistic understanding, where the predicative representation is fixed by the relatively ignorant questioner, that analysis is prematurely blocked.

We can thus see why the knowledge situation of the explanation seeker can be so different from that of the prediction seeker. By virtue of sometimes having the explanandum phenomenon itself available (materially, not just linguistically) operations of analysis and other modes of causal investigation are possible in ways that cannot be carried out in its absence. Moreover, when you have the actual object or process available to you, the modes of

---

[22] Understanding why the cat is on the mat is not the same as understanding why the sentence "the cat is on the mat" is true. This is because the answer to material questions such as the first often involve causal antecedents that led up to the state described by the sentence, whereas at least some answers to the second question involve the truth conditions at the time the explanandum sentence is true.

[23] How to assign properties or entities to various levels is not straightforward. It is not sufficient for A to be a (proper) part of B for A to be at a lower level than B, because this would consign subareas of plane figures to a lower level than the figure itself. Nor is it necessary, for in some cases of emergent entities, the emergent and higher-level entities do not have the lower-level entities that generate them as constituents. (For examples, see Humphreys 1997.) For present purposes, we shall simply take the conventional scientific taxonomy of levels of entities for granted, without endorsing it, and take the level of a property as that of the first level of entities at which it is instantiated. (This means that many lower-level properties will be found at higher levels as well.)

[24] This puts methodological limitations on such widely used statistical techniques as subclassification, e.g., Rosenbaum and Rubin 1983, sec. 3.3.

possible analysis are determined by the entity itself and not by the often impoverished syntax.

Of course, what one finds by analysis depends upon the structure of the analysandum. Often, analysis is required simply to produce an explanandum and the explanation has to be found elsewhere. For example, in 1963, the astronomer Maarten Schmidt realized that previously unidentified spectral lines from quasar 3C 273 resembled a Balmer series from hydrogen, but one that was greatly shifted toward the red end of the spectrum.[25] This value of the red shift required the quasar to be located 10^9 light years from Earth. Coupled with its brightness, this indicated that it must be emitting massive amounts of radiation. The spectral analysis thus had led to a why-question that was eventually answered through the gravitational action of black holes. In this case, the analysis provided the basis for formulating the question, but did not itself provide the answer.

Besides the examples we have already seen, there is a related problem that occurs most often in the social sciences. Here the problem is not one of having available the carrier rather than the cause itself but of using extremely general characteristics in place of the genuine, more specific, causal factors. The almost ubiquitous variables of class, race, and gender, which appear in so many causal models in the social sciences, are usually placeholders for more specific characteristics of individuals. For example, "class" is a cluster term standing for a collection of less abstract properties such as speech characteristics, tastes in music, literature, and food, choice of leisure activities, mode of dress, occupational aspirations, and so on. In any given context, not all of these are likely to be explanatorily relevant to the outcome of interest. More generally, which of them are explanatorily relevant in sociology often turn out to constitute a different subclass then those that are explanatorily relevant in economics, for example.

13.9. Analysis Can Provide Unification

But why try to isolate what we usually call the real cause in these situations, the aspect of tobacco that is carcinogenic, the active ingredient in willow bark, the aspects of class that lead to social advancement, and so on? It is because those constituents often occur in other objects or situations and we want to know what those other contexts are. By staying at the level of the object or the process, rather than going down via analysis to the level of the active cause, we restrict the generality of our understanding, as well as the range of our possible interventions, manipulations, and synthetic constructions.

---

[25] See Schmidt 1963.

We have seen that the same causal factor can be present in a wide variety of objects or processes. Consider the case of acids. It was once common to characterize acids functionally, as substances that produced carbon dioxide when added to sodium or calcium carbonate, or that turned red phenolphthalein colorless. In contrast to these functional (or operational) definitions, the contemporary Lewis concept of an acid defines it as a species that can accept an electron pair.[26] Here we have a property that is possessed by acetic acid, by sulfuric acid, by hydrochloric acid, and by many others. By virtue of having isolated this common property of acids, it has been possible to unify a wide variety of functionally characterized substances in terms of a single common feature.

This kind of progress by isolating common factors is well known and I note it only because the explanatory accounts that emphasize unification have focused on synthetic syntactic methods, rather than on analytic material methods. The example of unification through finding a structure common to most substances previously classified as acids was largely the result of theoretical work, but some of the other examples we have seen in which a more general understanding has been achieved are clearly analytic in nature. It is worth exploring the relation of this kind of unification to a more controversial feature that is usually claimed to lead in the opposite direction—to disunification. This is the multiple realizability of most functionally characterized properties.

To arrive at the comparison, we need to be explicit about the ontology and logical form of the realizability relation. Ordinary usage is somewhat flexible about the entities between which the realizability relation holds, and many philosophical treatments are almost as casual. The general form of the realizability relation is that of X being realized by Y in Z. X can be a property (first or higher order), a state type, or an object type; Y can be a property, a state or state type, or an object or an object type; Z is always an object or an object type. Properties and state types are probably the most common subjects of realization: Thus pain from ischemia is realized in humans by a specific type of neurophysiological state whereas pain from joint damage is (presumably) realized in iguanas by a distinct type of state. In addition, although only in an elliptical sense as we shall see, object types can be realized: a seat is said to be realized by a box, by a tree stump, by a dining chair, and so on. Frequently, the relation is collapsed by leaving either Y or Z tacit. Thus, the property of serving as money can be realized by banknotes, by electronic transfers, by company store chits, by gold, and so on, but it is assumed that this realization takes place in societies with monetary economies, rather than, say, societies based on barter. Similarly, the example of being a seat must be implicitly relativized

---

[26] The two other contemporary concepts of an acid are the Bronsted-Lowry and the Arrhenius conceptions.

to a physical context, such as human societies on Earth, for wicker chairs do not usually function well as seats for elephants.[27] It is not necessary for the first relatum to be functionally characterized: a binomial stochastic process can be realized by the particular sequence HTHHTTTHHTH . . . in a coin tossing setup. There is one kind of case where realization has the form of a two-place relation only and that is when the realizer is an abstract entity, as when an abstract group is realized by the group of rotations on two-dimensional Euclidean space.[28]

The apparent looseness of ontology here can be reduced because in any case in which an object type is said to be realized or to be a realizer, reference to the object type can be replaced by reference to an associated property or properties. So, it is the property of being a chair that is realized by properties of a tree stump, for example. In contrast, the third relatum must always be an object or object type.

Because state types are special cases of properties, we then have that when X is realized by Y in Z, X and Y are properties, and Z is an object or object type. This means that an instance of the realized property will be an instance of the realizing property, the instantiator being Z.

Realizability has attracted interest chiefly in the realm of multiply realizable properties. Multiple realizability is often taken to be an argument against the reduction of the multiply realized property to lower-level properties. Now, whether or not a given property or other entity is multiply realizable is ordinarily a contingent matter.[29] But so is the antireductionism that is often argued to result from it, for multiple realizability often coexists with unification resulting from a common underlying ontology. Consider the case of doorstops. The property of being a doorstop is functionally characterized and it can be multiply realized by certain properties of bricks, of cast-iron pigs, of wooden wedges, of bean bags, and so on, where all these are tacitly assumed to be located in our gravitational field. Yet there is only one relevant causal property possessed by all doorstops, and that is the ability to exert a counteracting force of magnitude greater than that exerted by the door. The apparent triviality of the doorstop example conceals something of importance, which is that physics discovered an extremely general common factor that is necessarily found in all properly functioning doorstops and in virtue of separating it from the other properties of doorstops, allows us to understand why they function as they do. This common factor also reunifies the functional type "doorstop"

---

[27] Alternatively, we could insert the relativization into the functionally defined concept, as in "is a chair on Earth."

[28] Because the core usage of "realization" usually conveys a transition from a more abstract entity to a more concrete entity, this last example could be taken as a courtesy use.

[29] Kim 1993, 312–13 suggests that some writers, notably those who construe mental properties as second-order properties, take their multiple realizability as a conceptual fact.

and thus makes its multiple realizability of engineering interest only, and not of physical or of philosophical interest. For a less trivial case, we have already seen the unification that was achieved with the Lewis definition of acids in this way. The dual moral is that property analysis in such cases can provide both understanding and unification. It also serves as a reminder that the connection from multiple realizability to nonreduction is quite shaky and that any appeal to the supposed implausibility of reduction from the existence of multiple realizability in these cases forms, to my mind, an extremely weak general argument against reduction.[30]

13.10. Laws Are Not Necessary for Understanding

Consider an utterly simple example of understanding, whereby we understand a phenomenon by showing that it operates as a Poisson process.[31] Suppose we are willing to make these assumptions about the stochastic process:

(a) During a small interval of time, the probability of observing one event in that interval is proportional to the length of that interval.
(b) The probability of two or more events in a small interval is small and goes to zero rapidly as the length of the interval goes to zero.
(c) The number of events in a given time interval is independent of the number of arrivals in any disjoint interval.
(d) The number of events in a given interval depends only upon the length of the interval and not its location.

These are informal assumptions, but ones that can be given precise mathematical representations. From those representations it is possible to derive the exact form of the probability distribution covering the output from the process. Such models are of great generality, and can explain phenomena as varied as the rate of light bulb failures and the rate of arrival of cars at a remote border checkpoint. In addition, unlike some models that require severe idealization or simplification, there is good reason to believe that in many applications the Poisson model is true of the processes it represents.

---

[30] Unification and reduction are, of course, different things, but in the kinds of cases I have described with common causal factors, it is plausible to take them as equivalent.

[31] For a simple exposition of Poisson processes, see Barr and Zehna 1971. In Humphreys 1995, I describe in detail the construction of models based on the diffusion equation. As a deterministic model, that is closer to the heart of the deductive-nomological approach than is the statistical model described here. However, deductive-statistical explanations were in all essential ways similar to deductive-nomological explanations in Hempel's approach, and the present example has the benefit of being purely structural.

Within the Hempelian approach of nomic expectability, it was necessary to include at least one scientific law in the explanans, whether explaining a particular fact or a general law. In many cases of model construction, laws will enter into the construction process, but in the example just described it is not in the least plausible that any of the assumptions given count as laws. Although one might say that the individual outcomes are explained by virtue of falling under a law-like distribution, if the fact to be explained is that the system itself follows a Poisson distribution, that fact is not understood in virtue of its being the conclusion in a derivation within which at least one law plays an essential role. Furthermore, even if we consider the distribution that covers the process to itself be a law, that law is not itself understood in terms of further laws, as the D-N account of explaining laws holds, but in terms of structural facts about the system. This, in my view, quite conclusively demonstrates that neither scientific understanding nor scientific explanations require laws.

13.11. Partial Understanding and Truth

One of Hempel's four criteria of adequacy was that the members of the explanans all be true. I believe that Hempel was correct in insisting on the truth of the explanans, but we must accommodate the kind of incompleteness with which science is always faced. There are two objections to the truth requirement that are often made.[32] The first is that an adequate theory of explanation has to accommodate the fact that at least some scientific theories that we now regard as false, such as Newton's gravitational theory, still have explanatory power because they are close to the truth. The second objection is that if we insist that it is necessary for the members of the explanans to be true, it is possible that none of our current "explanations" are actually explanations at all. (This is the so-called pessimistic induction from the history of science.)

In response to the first objection, one simply has to be explicit about what one means by a theory being close to the truth, or being partially true.[33] There are many sophisticated theories of verisimilitude available, but one rather different way in which an explanation can be partially true is when it postulates that there are multiple causal influences on a phenomenon and the proposed explanation gets some of those factors right but others wrong. Here is one example. In the 1962 Cuban missile crisis, it was said for many years that the Soviets withdrew their missiles from Cuba because (a) Khrushchev had backed down under pressure from a United States quarantine on the island,

---

[32] I have heard the first from Gary Hardcastle and the second from Stuart Glennan, among others.

[33] Hempel's treatment of partial explanations in section 4.2 of Hempel 1965a does not deal with these issues.

(b) the United States had provided guarantees that no invasion of Cuba would take place, and (c) the United States had publicly agreed to remove the quarantine measures on Cuba in return for removal of the missiles. It later become known[34] that Khrushchev and Kennedy had a secret understanding, brokered by Robert Kennedy, for the U.S. to remove its obsolete Jupiter missiles from Turkey in exchange for removing the Cuban missile bases. (a) thus turned out to be false, but (b) and (c) are still true.[35] So the original explanation still has some explanatory power to it and analysis, by separating independent contributions to the outcome, allows one to see why certain explanations are partially true. It also allows us to retain parts of imperfect explanations rather than rejecting them outright.

In a rather different sense it is claimed that Newton's theory of celestial motion or of bodies falling under gravity would be no explanation at all under the truth requirement. Once we acknowledge that theories and explanations based upon them are composite entities we can see that we can gain understanding from explanations that contain falsehoods because they are correct in part. Take for example Copernicus's explanation of the retrograde motions of the superior planets. Copernicus's theory was false in many respects: it contained compound circular orbits, it misrepresented the size of the universe, it contained only seven planets, it placed the Sun at the center of the universe, and it used crystalline spheres. Yet the basic qualitative features of retrograde motion of the Superior planets have remained explained by the basic geometry of a Copernican arrangement. The Newtonian case can be viewed in the same way. It appeals to an entity, a universal gravitational force, that does not exist in the sense that many Newtonians said it did (instantaneous action-at-a-distance). However, it cannot be a general principle of understanding that if a theory or model makes essential use of a (type of) entity which is such that nothing similar to that exists within the system to be explained, then that account fails to provide understanding, for this would rule out both Copernicus's theory and Newton's theory, as well as many others as having provided partial understanding. What is needed is a detailed account of how theories and models with both correct and incorrect features provide understanding of different aspects of the world. This is a large task that requires us to separate parts of theories that are construed instrumentally from those that are meant to be taken realistically, the role that idealizations

---

[34] See e.g. Hilsman 1996, 129–30; Thompson 1992, 345–6. As with most historical reconstructions, the facts became known only gradually, and some aspects of the development of the crisis and its resolution are still uncertain. Hints of the secret deal were made in Robert Kennedy's memoirs.

[35] A matter of considerable interest, although one that is essentially counterfactual in form, is whether President Kennedy would have violated the guarantees against a second invasion of Cuba had he not been assassinated.

and abstractions play in models, how model refinement is conducted, and so on. But that is a task for another time.[36,37]

References

Achinstein, Peter. 1983. *The Nature of Explanation*. Oxford: Oxford University Press.

Arnauld, Antoine, and Pierre Nicole. (1683) 1996. *Logic or the Art of Thinking*. Translation of the fifth French edited by Jill Buroker. Cambridge: Cambridge University Press.

Barr, Donald, and Peter Zehna. 1971. *Probability*. Belmont, CA: Brooks/Cole Publishing.

Cummins, Robert. 1977. "Programs in the Explanation of Behavior." *Philosophy of Science* 44: 269–87.

Friedman, Michael. 1974. "Explanation and Scientific Understanding." *Journal of Philosophy* 71: 5–19.

Hempel, Carl. 1942. "The Function of General Laws in History." *Journal of Philosophy* 39: 35–48. Reprinted in Hempel 1965b.

——. 1948. "Studies in the Logic of Explanation." *Philosophy of Science* 15: 567–79. Reprinted in *Hempel 1965b*.

——. 1965a. "Aspects of Scientific Explanation." In Hempel 1965b, 331–496.

——. 1965b. *Aspects of Scientific Explanation and Other Essays in the Philosophy of Science*. New York: Free Press.

Hilsman, Roger. 1996. *The Cuban Missile Crisis*. Westport, CT: Praeger.

Humphreys, Paul. 1989. *The Chances of Explanation*. Princeton, NJ: Princeton University Press.

——. 1995. "Computational Science and Scientific Method." *Minds and Machines* 5: 499–512.

——. 1997. "How Properties Emerge." *Philosophy of Science* 64: 1–17.

Kim, Jaegwon. 1993. *Supervenience and Mind*. Cambridge: Cambridge University Press.

Kitcher, Philip. 1989. "Explanatory Unification and the Causal Structure of the World." In *Scientific Explanation*, edited by Philip Kitcher and Wesley C. Salmon, 410–505. Minneapolis: University of Minnesota Press.

Régis, Pierre-Sylvain. 1690. *Système de philosophie, contentant la logique, la métaphysique, la physique, et la morale*. Paris.

Rosenbaum, Paul, and Donald Rubin. 1983. "The Central Role of the Propensity Score in Observational Studies for Causal Effects." *Biometrika* 70: 41–55.

Salmon, Wesley C. 1971. *Statistical Explanation and Statistical Relevance*. Pittsburgh: University of Pittsburgh Press.

---

[36] For first steps toward an answer, see Wimsatt 1998 and Humphreys 1995.

[37] For helpful comments on this essay I am indebted to audiences at the California Institute of Technology and the University of California at Santa Cruz. I particularly benefited from critical responses to section 13.7 from Fiona Cowie, David Freedman, Christopher Hitchcock, Ric Otte, Eric Scerri, and James Woodward. Not all of them agree with my conclusions. Some of the material in this essay descends, in a much revised form, from an unpublished paper read at the University of Pittsburgh and Johns Hopkins University in the early 1990s. Comments from those audiences were also very helpful.

———. 1984. *Scientific Explanation and the Causal Structure of the World.* Princeton, NJ: Princeton University Press.

———. 1989. "Four Decades of Scientific Explanation." In *Scientific Explanation,* edited by Philip Kitcher and Wesley C. Salmon, 3–219. Minneapolis: University of Minnesota Press. Reprinted as *Four Decades of Scientific Explanation* (Minneapolis: University of Minnesota Press, 1989).

Schmidt, M. 1963. "3C 273: A Star-Like Object with Large Red-Shift." *Nature* 197: 1040ff.

Snow, John. (1855) 1936. "On the Mode of Communication of Cholera." In *Snow on Cholera.* New York: Commonwealth Fund.

Thompson, Robert. 1992. *The Missiles of October.* New York: Simon and Schuster.

Wimsatt, William C. 1998. "False Models as a Means to Truer Theories." In *Re-engineering Philosophy for Limited Beings: Piecewise Approximations to Reality,* edited by William C. Wimsatt, 94–132. Cambridge, MA: Harvard University Press.

Scientific Ontology and Speculative Ontology

A conflict has emerged in contemporary philosophy between two quite different ways of approaching ontology. Over the last few years a growing divide has emerged between the fields that are often called "analytic metaphysics" and "scientific metaphysics." Analytic metaphysics is characterized by the importance given to a priori methods and to conceptual analysis. A posteriori scientific results are at best a peripheral part of its evidence base. In contrast, scientific metaphysics looks to the results of contemporary science for guidance to the correct ontology.[1] This tension between scientific and analytic metaphysics has a long history. The logical empiricists tried to do away with metaphysics, failed, and metaphysics made a spectacular comeback. Its revival increased the potential for friction between the two areas, although the antipathy seems to be more visible to philosophers of science than it is to metaphysicians. Some philosophers of science have principled objections to metaphysics or to conceptual analysis, often for reasons stemming from a sympathy for empiricism or for naturalism.[2] Others object to the use of nonperceptual intuitions as a reliable method for arriving at truths.[3] Still others find the pretensions of analytic metaphysics objectionable.[4] My paper is an attempt to lay out some of the reasons behind the tension and to suggest the form that appropriate ontological inquiries might take. There is a legitimate place for metaphysical reasoning

---

[1] Although scientific metaphysics tends to draw its conclusions from fundamental physics, I do not presuppose the correctness of that approach here. One reason is the failure of various reductionist programs and the empirical evidence for emergent properties that are not fundamental in the sense that they fall outside the domain of high-energy physics, digital physics, information-theoretic physics, or whatever is currently considered to be "fundamental."

[2] See, e.g., van Fraassen 2004 for an example of the first orientation.

[3] See, e.g., Williamson 2005.

[4] See Ladyman and Ross 2007.

and conceptual analysis in the philosophy of science. It is also true that parts of contemporary metaphysics are indefensible.[5]

Put in its simplest terms, the question is this: If one has reason to think that broadly aimed antimetaphysical arguments have failed and that metaphysics can be pursued as a supplement to science, what methods should a scientifically informed metaphysics employ? As a starting point, I shall assume that some version of realism is acceptable, where by "realism" I mean any ontological position that accepts the existence of at least some entities such that those entities and their intrinsic and relational properties do not depend on conscious intellectual activity.[6] This allows that some entities are so dependent, such as the property of fear, that some are culturally dependent, such as the existence of money, and that our own access to them may be mediated, perhaps unavoidably, by representational apparatus. The realism assumption is not unassailable, but there are few idealists nowadays, neo-Kantians can interpret the arguments given here as establishing choice criteria for the representational and evidential apparatus to be employed, and it allows many entities of traditional metaphysics, such as abstract objects, causal relations, haecceities, and others to be considered under the definition of realism.

This general characterization of realism is important for two reasons. First, more specific kinds of realism are defined in opposition to some other position, such as empiricism in the case of scientific realism, nominalism in the case of realism about universals, or constructivism in the case of realism about mathematical objects. One cannot ignore those oppositions entirely because they have to a large extent shaped the tradition of what counts as metaphysics, and it is the lack of clear success for the antirealist member of each of those pairs that has left open the possibility for the broader kind of realism considered here. But those oppositions are not directly relevant to my concerns. For example, the differences between scientific metaphysics and analytic metaphysics need not be due to differences about the status of abstract entities, as the older opposition between realism and nominalism suggested. Scientific metaphysics has to consider the possibility of abstract entities because first-order logic is too weak to serve as the basis for mathematics that is adequate for science and at least some second-order mathematical properties such as "is a random variable" are, if not purely formal, abstract.[7] Second, the kind

---

[5] In this article I shall discuss only selected examples of mainstream analytic metaphysics. There are more extreme forms but they are easier targets of criticism. In some cases the defense of a religious position appears to be the ultimate goal of the arguments.

[6] One thing that is not allowed to exist: conscious intellectual activity by a deity. This will be a restriction only if the existence of such a deity is used to transform some version of what is ordinarily considered to be idealism into a deity-dependent form of realism.

[7] The nominalist program for scientific mathematics pursued by Field (1980), while revealing, is not widely regarded as successful.

of broad ontological categories that emerge from these oppositions—abstract versus concrete entities, observable versus nonobservable entities, and so on— are too crude to be methodologically useful. The kinds of procedures that are relevant to establishing the existence of a set with the cardinality of the real numbers are radically different from those that would be needed to establish the existence of nonnatural moral properties.

Although the tension between the philosophy of science and analytic metaphysics is prima facie about what exists, that is proxy for a deep difference about what counts as an appropriate philosophical method. This means that the epistemology of ontological claims will play a central role in the discussion. I thus have a second assumption. It is that the methods of the physical sciences have been successful in discovering the existence and properties of a number of generally accepted entities in the physical sciences.[8] These procedures include empirical, mathematical, and computational methods. This assumption should also be reasonably uncontroversial and it does not deny that other methods can and have been successful when applied to appropriate subject matter. The assumption is incompatible with versions of universal skepticism under which all methods of inquiry are equally unable to provide warrants for truth but on those views the basic dispute under discussion here has no content. Weaker positions such that specifiable scientific methods are less fallible than the methods of analytic metaphysics could be substituted for our second assumption.

The two assumptions together provide us with an overall strategy. Realism suggests that because things are the way they are independently of how we know (of) them, and there is no prima facie reason to hold that a single, uniform set of epistemological methods will be successful in all domains, we should begin with the position that different domains of inquiry might require different epistemological approaches. Our second assumption suggests that we begin with those methods that have evidence in favor of their success in a given area and explore how far they can be extended before they break down and need supplementation. This suggestion does not preclude a priori methods and appeals to intuition being successful in areas such as mathematics and metaphysics. It relies on the inductive fact—and it is a fact—that those methods are unreliable when applied to the domain of the sciences. That is, we already have evidence that such methods fail when they are applied to the sciences mentioned above unless they are supplemented by empirical data. The project now is to see if incursions into metaphysics by scientific methods can be successful. It would be naive to simply assume that methods which

---

[8] By "physical sciences" I mean astrophysics (including cosmology and cosmogony), physics, and chemistry. I would be willing to extend this to some, but not all, biological sciences but this is not necessary for present purposes.

have been successful in the physical sciences will also be successful for ontological purposes in other areas, but a good deal of sound epistemic practice is enshrined in the methods of the more rigorous sciences and it can be useful to see whether they can be adapted to philosophical applications. If and when these methods break down, their failures can provide clues as to what should replace them in the given domain.

Having set up the broad opposition in these terms, I shall immediately restrict the scope of my discussion. It is impossible to precisely capture the scope of analytic metaphysics, which allows some practitioners to deny that they contribute to the discipline. So rather than discussing the broader category of "analytic metaphysics," I shall consider what I call "speculative ontology."[9] Consider ontology to be the study of what exists, a domain that includes not only what there is and what there is not, but the intrinsic properties of and relational structures between the entities, including second-order and higher properties and relations.

What restrictions does science place upon ontological results? I do not have a sharp criterion to distinguish scientific results from other kinds of conclusions. One reason is that considerable parts of contemporary cosmogony are so remote from direct empirical tests that they cannot be assessed primarily on grounds of empirical adequacy and are, in their own way, as speculative as are parts of metaphysics. But such activities are constrained by well-established scientific results such as general relativistic accounts of gravity, conservation principles derived from high-energy physics, and the Standard Model, together with widely accepted standards of scientific evidence such as consistency with empirical knowledge, use of the principle of total evidence, explicitness of the assumptions used, willingness to abandon a hypothesis in the face of counterevidence, and so on. All of these constraints are defeasible but they must be respected by those wishing to take into account all of the relevant contemporary evidence. The relevant differences between scientific ontology and other kinds of ontologies thus tend to center on what are considered constraints on the truth of the basic assumptions and the evidence base and when the kinds of constraints just mentioned are applied, we have scientific ontology. When they are not, we have speculative ontology.

Scientific ontology goes beyond whatever ontology the current scientific community currently endorses because underdetermination considerations often suggest the exploration of alternative ontologies that are compatible

---

[9] The term "speculative metaphysics" is often used to designate a certain type of activity, including the kind of metaphysics to which Kant was opposed. I shall therefore avoid that term. The term "naive ontology" is not quite right because the lack of appeal to scientific knowledge can be willful, although the effects are the same as would result from ignorance. The term "speculative mathematics" was traditionally used to denote what is now known as pure mathematics (see Stewart et al. 1835), a tradition not to be confused with how speculative mathematics is now conceived (see Jaffe and Quinn 1993).

with scientific principles and evidence and in which scientists display little interest.[10] Two examples are Bohm's interpretation of classical quantum mechanics (Bohm 1952) and David Chalmers's conclusions about dualism (Chalmers 1997). These are serious attempts to argue for an alternative to the mainstream ontology and both satisfy the conditions for a scientific ontology. I shall now present four principal reasons why speculative ontology should be viewed with deep suspicion. These are its widespread factual falsity, its appeal to intuitions, its informal use of conceptual analysis, and its assumption of scale invariance. These objections are not independent but because they do not apply uniformly to all activities within speculative ontology it is worth considering them separately.

14.1. The Factual Falsity of Speculative Ontology

Some claims of speculative ontology can be dismissed on the straightforward grounds that they are not even true, let alone necessarily true. The fault is not that these claims are false. Falsity is a ground for criticizing a claim, but not for dismissing it as irresponsible. What moves a claim from the first to the second category is the existence of an epistemic situation consisting of these components: (1) the claim is incompatible with the results of a well-established knowledge base, (2) those results are widely known in the relevant epistemic community, and (3) a modest amount of work would lead to those results being understood by philosophers. A murkier realm looms when the fact that the claim is false is brought to the attention of those who argue for it and the fact is ignored.

Traditional metaphysics tended to concern itself with necessary truths and although they were truths of our world too, the claims were generally beyond the reach of empirical science and so any potential conflict with scientific results was muted. Two things have changed that situation. The increased sophistication of the physical sciences has led to a number of metaphysical theses that were widely held to be necessarily true being shown to be factually false by experimental evidence. Well-known historical examples are the identity of indiscernibles, universal determinism, and universal causation.

Conversely, there has been a move by some metaphysicians toward taking certain metaphysical claims as contingent, rather than necessary. A much-discussed contemporary example of this kind of project is David Lewis's Humean supervenience program, the physicalist position within which the ontology consists only of local points of space-time and local properties

---

[10] Thus the deferential attitude to science endorsed in the natural ontological attitude (Fine 1984) is inadequate for an informed scientific ontology. I shall say more on this point in section 14.7.

instantiated at those points. Everything else globally supervenes on that. Here is what Lewis had to say about the modal status of his view: "I have conceded that Humean supervenience is a contingent, therefore an empirical, issue. Then why should I, as a philosopher rather than a physics fan, care about it? . . . Really, what I uphold is not so much the truth of Humean supervenience as the *tenability* of it. If physics itself were to teach me that it is false, I wouldn't grieve" (Lewis 1986, xi).

As an exercise in theorizing, Lewis's attitude would be unobjectionable were the position seriously put to an empirical test. But it has not in the sense that science long ago showed that Humean supervenience is factually false. The claim that the physical world has a form that fits the constraints of Humean supervenience is incompatible with well-known and well-confirmed theoretical knowledge about the nonseparability of fermions. It was empirically established before Lewis's position was developed that entangled states in quantum mechanics exist and do not supervene on what would in classical cases be called the states of the components. This feature of our world is sufficiently well confirmed as to make Humean supervenience untenable.[11]

It would be unreasonable to object to an intellectual activity simply on the grounds that its products did not represent truths about the concrete world. Many areas of mathematics have their own intrinsic pleasures and there is virtue in understanding how the world might have been, but is not. Yet when reading many Lewisian disciples, they do little to dispel the impression that the apparatus under construction is a guide to the ontological structure of our world and one wonders why so much effort is expended on this apparatus when more promising ontological alternatives are waiting to be investigated.[12]

Speculative ontologists might object to this line of reasoning on the grounds that what I have called factual falsity is better described as an inconsistency between a sentence that is supported by empirical evidence and a sentence supported by a priori reasoning, that the truth value of the scientific sentence is, on inductive historical grounds, uncertain and liable to eventually be rejected and hence that we should not allow the empirical evidence for the scientific claim to take priority over the a priori reasons for the other claim.[13] We can reply to this concern in two ways. First, that worries stemming from the pessimistic metainduction have been exaggerated; that raising the mere logical possibility that a scientific claim might be false adds nothing to what we already knew, which is that such claims are contingent; and that in the absence of specific evidence that is directly relevant to a particular scientific claim which suggests that the claim is false and the presence of specific evidence that it is

---

[11] See, e.g., Maudlin 2007, chap. 2.
[12] For a defense of Lewis's tenability claim, see Weatherson 2010.
[13] I am grateful to Anjan Chakravartty for raising this objection.

true, the rational epistemic action is to accept the claim as true. Such worries are more pressing when claims about fundamental ontology are involved and this response does indeed gain purchase there. But many conflicts between scientific and speculative ontology are not about fundamentals. Despite initial appearances, Lewis's metaphysics is as much about preserving common sense as it is about establishing esoterica.[14]

The second reply is to accept the response and to note that what science can do is to provide us with descriptions of possible situations that, even if they are not true of our world, expand our set of known metaphysical possibilities and can also show that certain metaphysical generalizations are not necessarily true. It is often as important to be aware of error as it is to know the truth and science is capable of informing us about both, within the limits of fallibility that are inescapable in both science and ontology.

14.2. Whose Intuitions?

A second source of discontent with speculative ontology is its appeal to intuitions as evidence. Although this suspicion is well-founded in many cases, arguments against the use of intuitions are not straightforward. Empirical evidence has shown that intuitions are highly fallible about both factual matters and reasoning, violating our principle of least epistemic risk (see section 14.6), but that evidence usually comes from studies on agents with skills below the expert level and we should be wary of too hasty conclusions about the inappropriateness of all appeals to intuition.[15] I take an intuition to be an unreflective and noninferential judgment and I shall be concerned primarily, although not exclusively, with nonperceptual intuitions.[16] It is common to add that intuitions are immediate, but insisting on immediacy qua temporal immediacy is unnecessary, for what is important is the unreflective aspect. Were I to be asked for my judgment on some issue, then immediately fell into a prolonged state of unconsciousness within which no subconscious or unconscious reasoning processes occurred, and I finally rendered my judgment upon awakening, or

---

[14] For a defense of this claim, see Symons 2008.

[15] In mathematics, views are divided about the use of intuitions. There is a stark contrast between the position of the logician J. Barkley Rosser, "The average mathematician should not forget that intuition is the final authority" (quoted in De Millo, Lipton, and Perlis 1979), and that of Frege, "To prevent anything intuitive from penetrating [the realm of arithmetical reasoning] unnoticed, I had to bend every effort to keep the chain of inferences free from gaps" (Frege, *Begriffsschrift* in van Heijenoort 1967).

[16] One use of the term "intuition" ruled out by this definition is its use in the expression "Here's the intuition," often said by mathematicians and other practitioners of formal methods after presenting some technical item. What is being conveyed is an informal understanding, stripped of necessary but secondary details, and that understanding is the result of significant reflection on the content of the formal apparatus.

the intervening time was spent mentally rerunning the last six miles of a recent marathon, my response would be as much an intuition as any instantaneous judgment. There are other uses of the term "intuition" in contemporary philosophy. In addition to the sense just given, it is common to use the term to refer to a psychological entity, a propositional attitude accompanied by a felt sense of certainty.[17] Brief reflection on the proposition involved is also allowed. Sometimes an intuition can be the starting point for a process of reflective equilibrium, but more often it is a nonnegotiable item.

Two obvious problems with intuitions are that they tend to differ, often considerably, between philosophers and they are often wrong when they are not about everyday experiences because we have no prior knowledge base on which to draw. These are familiar objections but I want to draw attention to a further reason to doubt many appeals to intuition. In traditions in which logical analysis and explicit definitions are central, it is often taken for granted that conceptual analysis and the use of intuition are subject-matter-independent activities and are largely agent-independent as befits their foundational status. But there are good reasons to doubt both of these claims and to hold that the reliability of intuitions and the use of conceptual analysis is domain specific and varies between agents. That is, a given agent's intuitions can be a reliable source of knowledge when applied to one domain and unreliable when applied to another while another agent's intuitions have the inverse degrees of reliability.[18]

This claim is clearly true for perceptual intuitions. Certain qualia are epistemically accessible by the visual modality but not others, such as flavors which must be accessed by the faculty of human taste. Other domains, such as those of mathematics and of modal truths, are assumed to be accessible by rational intuition or spatial intuition but are beyond the reach of olfactory intuitions.[19] Different agents tend to have different degrees of acuity, and so an idealization is made to ideally competent agents. In the perceptual realm, most humans were considered to have capacities sufficiently close to the ideal standard to enable them to come to know most observational truths. This was also thought

---

[17] For this use, see Bealer (1999, 247). Bealer argues that one can have an intuition that A and not believe that A because certain mathematical falsehoods can seem to be true yet one knows that they are not. This is arguable—if you do not believe A, you usually do not have the full force of a traditional intuition. It is not enough just to "seem" that A is true otherwise a guess will count as an intuition. I doubt that there is a uniform answer here, in part because different philosophers use the term "intuition" differently. Note that the modifier "rational" can be applied to intuitions, for example when extrinsic constraints such as consistency with other intuitions or Bayesian coherence criteria are imposed. For further discussions of intuitions, see DePaul and Ramsey 1998.

[18] Reliability is also agent-specific in that different agents have different degrees of reliability in a given domain, a fact that is often ignored in the exchange of intuitions in philosophical arguments.

[19] This surely is not necessarily true—there could be an arithmetic of smells—but I know of no mathematician who has reported working in this way.

to be true in the realm of a priori truths for ideally rational agents, and in that case the vast majority of humans were considered able, if not willing, to attain the standards needed for simple cases of a priori knowledge. Yet this level of generality is implausible.

Let us separate two claims. The first is that for a given agent, appeals to a priori intuitions have equal validity across all domains in which nonempirical knowledge might be available. This is demonstrably false. Intuitions about probability are unreliable for most people, one famous example being the Monty Hall problem.[20] There is a correct answer to the problem but most people arrive at a wrong answer when they approach the problem intuitively and their intuitions must be corrected either by arguments or by real or simulated frequency data.[21] In contrast, intuitions about the transitivity of preferences or the correctness of modus ponens are, outside a few problem cases, generally correct.[22] The second claim is that most agents are sufficiently close to ideally competent across all domains when using intuitions. This is also demonstrably false. When considering geometrical intuitions, number-theoretic intuitions, probabilistic intuitions, moral intuitions, logical intuitions, and so on, a given individual often has strengths in one area and not in another. So just as there are different perceptual modalities for different observable properties, and different scientific instruments for different detectable properties, the evidence seems to suggest that there are different levels of success for different kinds of intuitions, and that we must specify a domain of reliability for a given type of intuition and a given agent.

Of course, there have been attempts to reduce expertise in these disparate domains to a common basis. But in cases in which intuitions involve concepts, the burden of proof is on those who claim that all mature agents have equal reliability when appealing to intuitions, regardless of the concepts involved. This is at odds with what we know from practice. Some philosophers have better moral intuitions or logical intuitions than others, where the basis for evaluating reliability is long-term consistency between an individual's intuitive judgments and evidence gained from nonintuitive sources. This conforms to the widely held view in science and mathematics that physical, biological, mathematical, and other types of intuition come in degrees and there is no obvious reason why this should be different in philosophy. Furthermore, because

---

[20] For an extensive discussion of this problem see Rosenhouse 2009.

[21] Paul Erdős was initially certain that the accepted solution was incorrect and was convinced only after seeing the results of a computer simulation.

[22] Anyone who has taught logic to enough students will have encountered the curious phenomenon of an occasional student who seems to lack the ability to understand some primitive logical inferences. Sometimes they function using concrete inferences; in other cases one suspects that associations between categorical statements substitute for hypothetical reasoning, for example substituting the sequence "A customer complained. Call the supervisor" for inferences based on the rule "If a customer complains, call the supervisor."

in scientific areas the expert "intuitions" are usually, although not invariantly, shaped by considerable knowledge of the relevant science, the noninferential characteristic of intuitions will hold only at the level of conscious processing and this will also be characteristic of philosophical intuitions.

Behind the tacit assumption that the use of intuition is the basis of a general method often lies an appeal, inherited from the tradition of linguistic analysis, to competent users of a language. But semantic externalism and the division of linguistic labor entail that this competence, and the use of linguistic intuitions in conceptual analysis, has severe limitations.[23] If the semantic content of some part of language depends upon the way the world is, then knowledge of the world, often quite detailed knowledge of the world, is required to know what constitutes that content. Perhaps this is a minor difficulty since the use of intuitions is often restricted to core examples of the correct use of the expression.[24] Even so, there is a related and more serious problem that occurs independently of whether one subscribes to externalism and it again requires us to decide whose intuitions should prevail. Consider the extension of the predicate "dangerous." Applied to other humans, many of us have some reasonably reliable intuitions about whether a given individual has the associated property or not and can identify some core examples of dangerous humans. Yet we would, or should, defer to those who have better intuitions, such as members of the fire department, war veterans, or members of mountain rescue teams when circumstances demand it. What are the analogous circumstances in ontology and who are those with superior intuitions? I place little credence in appeals to intuitions and so I pose this challenge to the ontologists who use them. How do you train philosophers to improve their intuitions and how can we recognize when you have been successful? Who granted the Doctor of Intuition diploma on the metaphysician's mental wall?

We have arrived at the following principle: All methods have their domains of successful application and associated probabilities of error. There is not a uniform method of "appeal to intuition" that has the same success rate for all rational agents across all domains. In the presence of conflicts between intuitions and in the absence of criteria establishing who are the expert practitioners in a given domain, we should remain agnostic about any appeals to intuition and reject them as a source of evidence.

---

[23] Arguments for semantic externalism can be found in Putnam 1974.

[24] Although not always because considerable philosophical discussion occurs about borderline cases of concepts.

14.3. Conceptual Analysis

Conceptual analysis has a legitimate role to play in scientific ontology. Providing explicit or implicit definitions for concepts is an essential part of scientific understanding and should be encouraged but concepts drawn from everyday experience are often the wrong ones to analyze, whether in scientific or metaphysical contexts. One reason is that everyday, psychologically grounded concepts usually do not have sharp, necessary, and sufficient conditions, so that when an explicit definition is set up, it will not fit all intuitive examples of the concept. This is often the basis for the kind of unproductive exchange well-known to those who attend philosophy colloquia. In scientific contexts, conceptual analysis is complicated by the use of approximations and idealizations and a great deal of care is needed to draw the distinction between an idealized theoretical concept and the concept that is used in applications. Despite these reasons for maintaining a clear distinction between refined, theoretically based concepts and concepts drawn from less considered sources, appeals to intuition have begun to serve as a replacement for traditional conceptual analysis in some areas of philosophy and even sophisticated philosophers with high standards for argumentation have suggested that their use is acceptable.

Take Frank Jackson's characterization of conceptual analysis: "conceptual analysis is the very business of addressing when and whether a story told in one vocabulary is made true by one told in some allegedly more fundamental vocabulary" (Jackson 2000, 28). This characterization covers not only familiar cases that lie in the vicinity of analytic truths and falsehoods, but the kinds of traditional reductive procedures in the philosophy of science that considered whether condensed matter physics is simply redescribing complex combinations of phenomena for which high-energy physics has its own language.[25] Presupposed in this account is the view that we can tell when we are describing the same things in different ways. When this decision procedure relies solely on our understanding of some language, appeals to a priori knowledge are reasonably unproblematical. But the situation is different in discussions of reduction because the higher-level concepts are often presented to us through a theory at the higher level and not through explicit definitions or a quasi-reductive compositional account of the ontology. The identification of the referents of the different descriptions then often requires empirical knowledge and cannot be carried out a priori, a fact that is reflected in necessity of identity claims resting on the contingent truth of the identity.

---

[25] The question of whether the Nagelian bridge laws used in that tradition are empirical or definitional is obviously relevant to whether conceptual analysis in this sense is sufficient for this example. The fact that neither the concepts of condensed matter physics nor its ontology are reducible to those of high-energy physics is but one reason why many mereological claims fall into the category of speculative ontology.

Furthermore, the evidence for the existence of the higher-level phenomenon is empirical, not a priori. For example, purely theoretical, ab initio derivations of high temperature superconductivity are infeasible and the approximations used to arrive at the higher-level theory are justified on experimental grounds, not on a priori reasoning and definitions.

Not all contemporary conceptual analysis takes the traditional form of definitions. For example, David Chalmers and Frank Jackson have argued that conceptual analysis based on explicit or implicit definitions is unavailable and unnecessary in many epistemological contexts. "It is sometimes claimed that for A ⊃ B to be a priori, the terms in B must be definable using the terms of A. On this view, a priori entailment requires definitions, or explicit conceptual analyses: that is, finite expressions in the relevant language that are a priori equivalent to the original terms, yielding counterexample-free analyses of those terms. This is not our view. . . . If anything, the moral of the Gettier discussion is that . . . explicit analyses are themselves dependent on a priori intuitions concerning specific cases, or equivalently, on a priori intuitions about certain conditionals. The Gettier literature shows repeatedly that explicit analyses are hostage to specific counterexamples, where these counterexamples involve a priori intuitions about hypothetical cases" (Chalmers and Jackson, 2001, 320–22). On this view, intuition plays a central role but since far more than a mastery of language is needed to bridge the levels in realistic reductionist examples, this position is less than convincing.

Goldman (2007, 18) argues that "Philosophical analysis is mainly interested in common concepts, ones that underpin our folk metaphysics, our folk epistemology, our folk ethics, and so forth." That would be fine, and indeed it may be the best approach for many issues in ethics, for ethics cannot move too far from what seems instinctively correct to most people without encountering overwhelming resistance. The situation is different in ontology because metaphysical arguments have led to highly counter intuitive positions. The fact of being counterintuitive is not itself an objection, for science frequently uses counterintuitive representations too, but these are usually supported by empirical data. The problems arise when metaphysicians venture judgments about domains for which the appropriate concepts are far removed from, and often alien to, our human intuitions. Humans are epistemic experts in certain limited domains of inquiry, those that are best suited to our naturally evolved cognitive apparatus, and this includes many ordinary moral judgments. Had we evolved in a different environment, it is likely that our concepts would have been different, probably very different. Bees are the product of evolution, but there is little reason to think that they have similar concepts to ours.[26] This

---

[26] There is empirical evidence that bees perceive the world differently from us and thus would be likely to have a bee ontology that is different from human ontology. See Chittka and Walker 2006.

suggests that when considering ontological issues, we should also consider cognitive agents who are not subject to our own contingent limitations and are different from or superior to us in cognitive abilities.

An example from computational science may illustrate why this appeal to extended cognitive abilities is useful in certain situations. We are delegating significant parts of computationally based science to nonhuman executors in computational neuroscience, complexity theory, condensed matter physics, and many other domains, a trend that is destined to accelerate. We shall then with increasing frequency require novel conceptual frameworks to, first, employ those techniques in the most effective possible ways and, second, to understand the results emerging from their deployment. This is because what makes an effective representation for a computer is frequently different from what we humans find transparent in the conceptual realm.

Consider how computerized tomography represents data. An X-ray beam is directed through an object and impinges on a detector which measures the intensity of the received X-rays. If $f(x,y)$ represents the intensity of the X-rays at the point $<x,y>$, then the line integral along the direction of the beam L, which is represented by the Radon transform of f along L, sums the intensities at all points along the line. Rotating the source around the object and plotting the values of the Radon transform for each of these angles gives a sinogram, which is what the detectors "see." It is rarely possible to know what the irradiated object is by visually inspecting the sinogram. Yet computers, using the inverse Radon transform, can easily produce data that, when turned into a graphic, "directly" represent the object, such as a human skull, in a form accessible to the human visual system.[27]

The sinogram image has no compact representation in human languages connecting it with the ordinary perceptual concept of a human skull and since the machines operate on purely extensional representations, this example directly raises the problem that intensional characterizations of properties in natural languages are likely to be too sparse to capture the kinds of extensionally characterized properties used by computational science. What constitutes a pattern for a human is more restricted than (or is just different from) what constitutes a pattern for a computer. This suggests that philosophers need to do one of three things; expand their repertoire of representations to include those used by computational devices and instruments, despite the fact that those representations do not conform to any that are intuitively accessible to humans; concede that certain current and future scientific activities lie beyond human understanding, assuming that understanding requires possessing the relevant intensional representations, a position that is at odds with the fact that

---

[27] For examples of sinograms see http://demonstrations.wolfram.com/ComputedTomographySimu lationUsingTheRadonTransform/ accessed May 23, 2012.

we can understand at least some of these activities; or push their explorations of subconceptual, nonrepresentational, and nonconceptual content beyond the realms currently explored in artificial intelligence, such as the use of dynamical systems theory or neural net representations. The second and third options entail that there are parts of ontology that lie beyond what is currently intuitively accessible, while the first option is unlikely to be successful without a considerable amount of conceptual retraining.

14.4. Philosophical Idealizations

Consider the following claim. Having insisted that a collective, long-term cognitive effort by humans allows an approximation to the ideal cognitive conditions required for a strong modal tie of intuitions to the truth, the author goes on to write that

> Some people might accept that the strong modal tie thesis about intuition . . . [is] non-empirical but hold that [it does] nothing to clarify the relation between science and philosophy as practiced by human beings. After all, these theses yield only the possibility of autonomous, authoritative philosophical knowledge on the part of creatures whose cognitive conditions are suitably good. What could this possibly have to do with the question of the relation between science and philosophy as actually practiced by us?

The answer is this: The investigation of the key concepts—intuition, evidence, concept possession—establish the possibility of autonomous, authoritative philosophical knowledge on the part of creatures in those ideal cognitive conditions. The same concepts, however, are essential to characterizing our own psychological and epistemic situation (and indeed those of any epistemic agent) (Bealer 1998, 202–3)

So what are these idealized cognitive conditions? In the article by Chalmers and Jackson cited in section 14.3, the authors write: "A priority concerns what is knowable in principle, not in practice and in assessing a priority, we idealize away from contingent cognitive limitations concerning memory, attention, reasoning, and the like" (Chalmers and Jackson 2001, 334). But this view quickly leads to idealizations that are questionable and commit us to epistemic abilities that can trivialize philosophical conclusions. In the realm of idealizing human abilities, if we were allowed to sufficiently idealize our limited abilities, an appeal to supernatural agents would always be available. Moreover, since it has been remarked that the correct metaphysics will be revealed at the limit of idealized epistemology, some, although not all, epistemological idealizations of the "in principle" kind will render certain claims of analytic metaphysics unassailable by default.

Although these approaches often go by the name of "in principle" approaches, they are more accurately seen as forms of idealization and this requires us to answer the question: What counts as a legitimate philosophical idealization? The analogous question for scientific domains has been much discussed, but there is little in the way of sharp answers to the philosophical question.[28] Just as it is appropriate to require of scientific idealizations that a given idealization maintains some contact with properties of real systems, we should similarly require that we have some criteria for how to relax philosophical idealizations to bring them into contact with human abilities. One kind of epistemic idealization involves the extrapolation of human abilities. We can perform arithmetical operations on integers of a certain size and we then extrapolate that ability to much larger integers that human memory capacity is incapable of storing. That kind of idealization seems to be philosophically legitimate for some purposes because we know the algorithm for generating integers, we are familiar with memorizing small collections of them, and extrapolating that familiar experience to larger collections is just "more of the same."[29] A different kind of epistemic extension involves augmentation. Examples of this are easy to give for sensory modalities. We are not naturally equipped to detect the spins of elementary particles and we require instruments the output of which must be converted into a form that we are equipped to detect. Here it is unreasonable to suggest that detecting spins involves an in-principle idealization of human abilities, for there is nothing like that ability in our perceptual repertoire.

The epistemology of speculative ontology relies heavily on cognitive extrapolations and augmentations; the appeal to the perfections of God in the ontological argument, an appeal that runs counter to the standard theological position that, lacking the full complement of those perfections themselves, it is blasphemous for humans to suggest that they can fully understand the concept of God; the appeal to truths at the limit of scientific inquiry, when we have no conception of what limit science will be like; oracles, or Platonists, who can inexplicably access mathematical truths, and so on.[30] Whether these are legitimate extrapolations or augmentations of human intellectual powers is an open question in the absence of criteria for acceptable philosophical idealizations.

---

[28] The literature on the differences between those who appeal to ideally rational agents and those who require bounded rationality is relevant here.

[29] This argument does not apply when we are accounting for how much of science is applied in practice.

[30] For clarification: the ontology of limit science here is included in the category of speculative ontology because there is no currently available way of inductively inferring from the present state of science to its limit state.

14.5. Scale Variance

Here is another principle: *Whether the world is scale-invariant is a contingent fact.* There is evidence that it is in certain ways and that it is not in others. Humans are all inescapably middle-sized objects with limited cognitive capacities that developed in the course of dealing with properties associated with similarly sized objects. One of the great scientific shifts in perspective was the twentieth-century realization that physical phenomena were not universally scale-invariant, that classical mechanics did not extend unchanged to the realms of the very small and the very large (nor, as a result, was it exactly correct in the realm of the middle-sized).[31] Once we had arrived at that realization, the status of the human a priori and of human experiences as sources of knowledge should have changed dramatically. Yet much work in epistemology and metaphysics proceeds as if that scientific shift had never occurred.

What properties, relations, objects, laws, and other parts of our ontology are like at scales radically different from those with which we are natively equipped to deal cannot be inferred from the evidence of direct experience. Consider this simplistic argument: ordinary experience tells us that a magnet has two opposite poles. If one cuts a bar magnet in two, each half has both a north and south pole. Therefore, by induction on common experience, one concludes that magnetic monopoles should not exist. And indeed, magnetic monopoles have not to this point been detected. But the scientific arguments for and against the existence of magnetic monopoles are of a very different kind than the simplistic argument just given and current versions of quantum field theory suggest the existence of dyons, of which magnetic monopoles are a special case. For a less naive inference, take the limit of zero resistance in a suitable conducting material (usually achieved by lowering the temperature) and you have moved into a distinctively different physical realm, one in which not just Ohm's law is false but the ontology changes as the conductive carriers, which originally are electrons, are replaced by a Cooper pair superfluid.

So what morals should we draw from the lack of scale invariance? First, the distinctions between the observable and the unobservable, between the nontheoretical and the theoretical, and between entities at the human scale and those outside that scale are different distinctions. The fact that the first two draw on different categories, objects, and properties in the first case, terms in the second, has often been noted. The logical independence of the elements of the second and the third distinctions can be shown by the fact that money is a

---

[31] I exclude here renormalization theories that focus on scale-invariant phenomena. Ladyman and Ross 2007 have discussed similar issues under the idea of "scale-relative ontology." Their discussion of Dennett 1991 on which their scale-relative position is based is very helpful in understanding the consequences of Dennett's paper.

classically human-scale construct but there exist multiple different theoretical measures of what falls under its scope—M0, M1, M2, and M3 are common measures. Also, the description "a sphere twice the size of the largest sphere considered to be at the human scale" is nontheoretical, whatever nontrivial sense of "theoretical" you use, but describes an entity beyond the human scale. The other two cases are easy. The fact that the third and the first distinctions are incommensurable becomes evident if we focus on properties rather than objects. Properties can be observable or not, but size scales do not apply to properties, including metric properties such as "is one foot long," only to regions of space possessing them.

Now consider the standard definition that induction involves inferences from the observed to the unobserved. Stock examples of inductive inference, such as generalizations from subpopulations to the whole population, which involve claims that the same property possessed by members of the subpopulation will be possessed by all members of the population, can be misleading because usually all instances of the generalization exist at the same scale. But inferences from the observed to the unobserved across differences of scale introduce an additional inductive risk. In addition to the usual assumptions such as that the future resembles the past and that properties are projectable, we must make the assumption that the part of the world under investigation is scale-invariant. This kind of inference is often classified as extrapolation, and extrapolation is, at root, a particular kind of inductive inference and is hence not a priori. Thus what might seem to be a result of a priori reasoning within speculative ontology, and this is especially true of various forms of mereology when put forward as a comprehensive ontological position, actually contains a hidden a posteriori element that is, moreover, likely to be false. This argument applies independently of what particular perceptual abilities an agent has, as long as they are not universal, and the concepts involved in the ontology require empirical access to be acquired.

Much of metaphysical argumentation has deep similarities with the arguments that accompany thought experiments and when constructing a hypothetical scenario, we need to know the laws and specific facts that apply at the scale at which the scenario is described. The only source of that knowledge is science; our intuitions and a priori imagination and reasoning are not equipped to provide it. At least some of the problematical examples of speculative ontology arise from simply assuming scale-invariance when the system does not exhibit it. For example, the telegraphic identity "water = H_2O" is false as stated since the left-hand side refers to the usual macroscopic fluid and so the right-hand side should read "a macroscopic collection of H_2O molecules interacting in such a way that the properties of liquidity, transparency, ability to undergo phase transitions, and so forth are present."[32]

---

[32] For a related point, see Johnson 1997.

Here then is a central problem of using intuitions in speculative ontology. The apparently a priori methods mask a tacit appeal to an inductive inference when making the scale-invariance assumption. Because human intuitions about ontology are obtained by experience with human-sized entities, any inference to regions beyond that domain involves an inference from the empirically known to the empirically unknown and that contains an inductive risk. Because there is considerable evidence that this inductive risk is high and that the conclusions of similar inferences have turned out to be false, generalization from intuitions in speculative ontology should be avoided.

14.6. The Principle of Least Epistemic Risk

One feature that traditional metaphysics and traditional empiricism had in common was a commitment to risk aversion. It can be captured in this principle: When competing ontological claims are made, determine the degree of epistemic risk associated with the methods used to establish or deny the existence of the entity in question and make the ontological choice based on the method with the lowest risk. The risk need not be zero—certainty may not be obtainable—but in their different ways, the foundational enterprises embedded in traditional metaphysics and traditional empiricism both relied on this principle, together with a second principle that inferences from the foundations never decrease the degree of risk, making the foundations the most secure of all claims.[33]

This epistemic principle has important consequences for ontological claims. For example, it makes the opposition to scientific realism by traditional empiricism inappropriate because traditional empiricism takes observables to constitute a fixed category tied to human perceptual abilities and it takes beliefs about observables to be uniformly less risky than beliefs based on what are considered to be unobservables.[34] I criticized the first assumption in Humphreys 2004, chapter 2; here I focus on the second assumption. Suppose we take the human perceptual apparatus as the foundation for all beliefs about concrete entities. In this role, the human perceptual apparatus plays the same function as calibration devices in the sciences. If we believe that a room measured with a laser rangefinder has a length of 22 feet, this is based on the user's belief that were he to use a tape measure instead, he would directly see the coincidence between the 22 feet mark on the tape and the wall of the room. Assessments of systematic and random errors for an empirical quantity are always made with respect to a calibration standard for the "true" value. But the

---

[33] Of course inferences together with additional evidence might result in a decrease.

[34] I have couched the discussion here in terms of beliefs to accommodate traditional empiricist positions although my own view is that this anthropocentric apparatus should be abandoned.

device used for the calibration standard is rarely the one used to collect data in most experiments because the conditions under which data from a particular system must be collected are not those under which the calibrating system has optimal functionality. To illustrate the different domains of applicability of traditional empiricism and scientific empiricism, at the time of its painting a portrait by Vermeer could have been authenticated by his wife who witnessed him painting it but now, X-ray fluorescence and wavelet decomposition techniques would supersede any human judgment. Differently, although the correctness of individual steps in a long computation can be determined a priori by humans, the balancing of foreign exchange trading by Deutsche Bank AG is made by automated methods, and unavoidably so. The philosophical moral is this: the fact that there exist agents who act as a ground for beliefs about a given epistemic task does not entail that those agents can act as a foundation for comparative judgments about risk for that task under all conditions. As science gains greater access to what exists, the judgments of humans, whether empiricists or speculative ontologists, become increasingly risky and so will be replaced by conclusions drawn from scientific sources.

14.7. Why Philosophers of Science Cannot Avoid Doing Metaphysics

Some of the problems of speculative ontology have been described. Does this mean that we should avoid ontological or metaphysical arguments entirely? This is not possible and to see why consider the following question. How did we get to the point where it is necessary to once again eliminate certain types of metaphysical reasoning? Here is one broad historical thread that gives at least part of the reason. By its very nature, much of metaphysical activity involves decisions about issues that are underdetermined by science. In particular, for decisions about ontology, this route was signaled by Carnap's characterization of external questions as having conventionally chosen answers and by Quine's relativized ontology. (See Carnap 1956 and Quine 1969.) Wanting such choices to be based on less vague grounds than the pragmatic criteria offered by those two philosophers, some metaphysicians, understandably, have tried to provide convincing arguments for the choice of one ontology over another.

This is a legitimate role for philosophy. Ordinary scientific practice is not oriented toward establishing claims of realism or antirealism. For the most part, when scientists try to do philosophy, they do it as amateurs, with noticeably poor results.[35] Science is designed to obtain results from complex

---

[35] There are exceptions. Richard Feynman, despite his often caustic comments about the philosophy of science, had a number of important philosophical insights, some contained in Feynman 1967. Other

experimental apparatuses, to run effective computer simulations, to derive predictions from models, along with other tasks unrelated to ontology. In arguing for his Natural Ontological Attitude in which one leaves the choice of ontology to the relevant scientific community, Arthur Fine suggested, in analogy with Hilbert's program in metamathematics, that "to argue for realism one must employ methods more stringent than those in ordinary scientific practice" (Fine 1984, 85–86). This analogy with consistency proofs is misplaced. The justification procedures involved in using observational or experimental techniques to assert the existence of some entity or type of entity are not stronger than the theoretical representations involved in making the realist claims, merely different.[36] Moreover, even in formal sciences, philosophical arguments about existence are relevant. Here is an example. Using a consistency proof, Herzberg (2010) has shown that infinite regresses of probabilistic justification do not render justification of the end-state proposition impossible but he notes that the value of the justificatory probability may not be expressible in the form of a closed-form solution. Because there is no precise, universally accepted, definition of a closed-form solution, reasons for and against the acceptability of probability values that cannot be estimated are not mathematical but recognizably philosophical, whether presented by philosophers or by mathematicians.

Some scientific activities, such as the search for the Higgs boson or the earlier search for the conjectured planet Vulcan, also have the aim of establishing the existence or nonexistence of certain features of the world and philosophers can learn much from the procedures used. There is a strong tradition in arguments for and against realism that criteria for what exists must be general. Yet one moral that can be drawn from claims of existence made within science is that the evidence for existence claims is often domain-specific. This suggests that a position of selective realism is the kind of realism appropriate for scientific realism, rather than the uniform commitment to realism that is characteristic of, to take one example, Quine's criterion for ontological commitment.[37] Coupled with Quine's resistance to second-order logic, the legacy of this criterion has distorted discussions of ontology. We can also be misled by too strong an emphasis on entity realism or statement realism, where entities are identified with

---

scientists, equally critical of the philosophy of science, are simply naive, which is what one would expect of those with no training in the field. For one example, see Hawking and Mlodinow 2010.

[36] "Observational" is used here in the liberal scientific sense, not in the philosophical sense.

[37] The "realism" that results from the appearance of existential quantifiers in a theory is only a genuine form of realism when objectual quantification over domains of real objects is used. Quantification over domains of models, or other kinds of weaker interpretations of quantification are insufficient to establish genuine realism. Of course, using domains of real objects is, without clarification, blatantly circular. For an account of selective realism see Humphreys 2004, sec. 3.8.

objects, when an important part of the realism issue involves the existence of properties.

These are considerations relating to scientific realism and although empiricists and antirealists have attempted to label specifically philosophical reasoning about such matters as illegitimate, such reasoning is to specifically ontological conclusions. Here is one further example that is indisputably metaphysical. Universal determinism can seem puzzling because it is committed to the view that the present state of the universe was fixed millions of years ago. The puzzlement can be made to go away, but there is a related puzzle that is less easy to resolve: How can all of the relevant information needed to fix future states of the universe be encoded into a single instantaneous state? This puzzle becomes more pressing when we consider the state at the first instant of the universe. How can everything needed to guide the development of the universe for evermore be right there in a single time slice which has no history and no law-like regularities to guide its subsequent development? This is a serious difficulty for any view that insists that there can be no causal properties without an associated law, and that regularities are a necessary condition of having a law. There are no nontrivial regularities at the start of the universe, hence no laws, and hence, on these views, no causes.

There is no difficulty for systems that can refer to rules. Computational systems are such that their dynamical development is driven by rules of the form: if the system is in such and such a state, then execute action thus and so. If we assume analogously that the laws of nature were there at the beginning, as part of the program for the universe, this answer immediately gives us an important realist conclusion, that the early stages of the universe contained laws that existed independently of specific states and of what we can know about them.

There are other approaches to addressing this puzzle. One says simply that there is no answer to our question, that there are just successive states of the universe, one after another, and parts of the first time slice can be attributed a causal role only if and when regularities appear. I note that this answer can be, and is, couched in realist terms rather than in the epistemic way that Humeans favor. If you feel that there are deep and probably unanswerable mysteries about the origin of the universe, then you should feel that this first approach attributes those mysteries to every successive instant in the universe's evolution until a satisfactory regularity had been established. This is Hume's Problem dramatized, and it leaves the early and subsequent development of the universe a complete mystery. It is consistent but lacking in explanatory content.

A second approach denies the implicit assumption of the puzzle, which is that at the first instant there are no later instants yet in existence, and adopts the block universe model with its assumption that all states are eternally coexistent. This second approach has a curious feature, for if in such worlds laws supervene upon regularities and at least some different subsequent histories of the universe would give rise to different laws, and explanations of why later states of the universe have the features that they do depend upon both the initial state and the laws, then it is not true that the initial state of the universe determined (by itself) what was to come. Nor indeed does any initial segment of the universe, but only its entire subsequent development. This involves both the dependence of earlier states on later states and action at a temporal distance. A high price to pay and one that can be avoided by a third approach, which is to reject the idea that laws play a role in the development of the universe, and to assert that it is present property instances and their interactions that alone give rise to future developments. This is a metaphysical argument and one to which science itself has little if anything to contribute beyond providing constraints on the possible solutions.

14.8. Conclusions

The source of many of the problems associated with speculative ontology is that science, and physics in particular, long ago outran the conceptual abilities of most speculative ontologists. Contemporary science has revealed a much more subtle and interesting world than the often simple worlds of speculative ontologists, one example being the overuse of mereology by many contemporary metaphysicians.[38] The solution to these inadequacies is to pursue scientific ontology, an activity that is primarily interested in the contingent ontology of our world but one that can also provide a guide to unactualized possibilities. We can extract six morals from our discussion.

First, metaphysicians could, with a modest amount of effort, identify which contemporary metaphysical claims have been shown by science to be either contingently, rather than necessarily, true or simply false, thus preventing a considerable amount of wasted effort. It is true that advocates of speculative ontology can point to the pessimistic induction from the history of science as evidence that all scientific knowledge is eventually overthrown and argue that appeals to scientific authority are therefore too temporary and flimsy a basis for metaphysical truth. This would be an acceptable response if appropriate controls are placed on the methods used in metaphysics. A scientific ontologist must concede that fallibilism is the only acceptable epistemological position but our second conclusion, that different methods are appropriate for different domains of investigation, should be implemented. Third, appeals to intuition are to be viewed with suspicion and it must at least be

---

[38] There are scientifically sensitive advocates of more sophisticated forms of mereology. For one example, see Arntzenius and Hawthorne 2005 and for a balanced assessment of the mereological project see Varzi 2011.

recognized that such appeals are both domain sensitive and differ in reliability between philosophers. Fourth, conceptual analysis needs to rely less on ordinary concepts and draw more heavily on the extensive conceptual resources of science and, where appropriate, mathematics. Fifth, philosophers collectively need to develop criteria for what counts as a legitimate philosophical idealization, criteria that are currently conspicuous by their absence. Finally, ontologists must recognize that our world is not scale-invariant and that as a result, a number of inferences that are taken to be a priori have a hidden inductive element.[39]

References

Arntzenius, Frank, and John Hawthorne. 2005. "Gunk and Continuous Variation." *The Monist* 88: 441–65.

Bealer, George. 1998. "Intuition and the Autonomy of Philosophy." In *Rethinking Intuition*, edited by Michael DePaul and William Ramsey, 201–39. Lanham, MD: Rowman and Littlefield.

———. 1999. "The A Priori." In *The Blackwell Guide to Epistemology*, edited by John Greco and Ernest Sosa, 243–70. Oxford: Blackwell.

Bohm, David. 1952. "A Suggested Interpretation of the Quantum Theory in Terms of 'Hidden Variables.'" Parts I and II. *Physical Review* 89: 166–93.

Carnap, Rudolf. 1956. "Empiricism, Semantics, and Ontology." In *Meaning and Necessity*, 205–21. Chicago: University of Chicago Press.

Chalmers, David. 1997. *The Conscious Mind*. Oxford: Oxford University Press.

Chalmers, David, and Frank Jackson. 2001. "Conceptual Analysis and Reductive Explanation." *Philosophical Review* 110: 315–61.

Chittka, Lars, and Julian Walker. 2006. "Do Bees Like Van Gogh's Sunflowers?" *Optics & Laser Technology* 38: 323–28.

De Millo, Richard A., Richard J. Lipton, and Alan Perlis. 1979. "Social Processes and Proofs of Theorems and Programs." *Communications of the ACM* 22: 271–80.

DePaul, Michael, and William Ramsey, eds. 1998. *Rethinking Intuition*. Lanham, MD: Rowman and Littlefield.

Dennett, Daniel. 1991. "Real Patterns." *Journal of Philosophy* 88: 27–51.

Feynman, Richard. 1967. *The Character of Physical Law*. Cambridge, MA: MIT Press.

Field, Hartry. 1980. *Science without Numbers*. Princeton, NJ: Princeton University Press.

Fine, Arthur. 1984. "The Natural Ontological Attitude." In *Scientific Realism*, edited by Jarrett Leplin, 83–107. Berkeley: University of California Press.

---

[39] I should like to thank the many philosophers with whom I have discussed the topics covered in this paper, especially the participants in a seminar at the University of Virginia in the spring of 2007, the audience at the University of Alabama conference on scientific naturalism and metaphysics in the fall of 2009, and lectures at the Ecole Normale Supérieure, Paris, the American Philosophical Association Pacific Division, and the University of Valparaiso, Chile. Comments on the penultimate draft from Anjan Chakravartty were particularly useful.

Goldman, Alvin. 2007. "Philosophical Intuitions: Their Target, Their Source, and Their Epistemic Status." *Grazer Philosophische Studien* 74: 1–26.

Hawking, Stephen, and Leonard Mlodinow. 2010. *The Grand Design*. London: Bantam Books.

Herzberg, Frederik. 2010. "The Consistency of Probabilistic Regresses. A Reply to Jeanne Peijnenburg and David Atkinson." *Studia Logica* 94: 331–45.

Humphreys, Paul. 2004. *Extending Ourselves: Computational Science, Empiricism, and Scientific Method*. New York: Oxford University Press.

Jackson, Frank. 2000. *From Metaphysics to Ethics: A Defence of Conceptual Analysis*. Oxford: Oxford University Press.

Jaffe, Arthur, and Frank Quinn. 1993. "'Theoretical Mathematics': Towards a Cultural Synthesis of Mathematics and Theoretical Physics." *Bulletin (N.S.) of the American Mathematical Society* 29: 1–13.

Johnson, M. 1997. "Manifest Kinds." *Journal of Philosophy* 94: 564–83.

Ladyman, James, and Don Ross. 2007. *Every Thing Must Go: Metaphysics Naturalized*. Oxford: Oxford University Press.

Lewis, David. 1986. *Philosophical Papers*. Vol. 2. Oxford: Oxford University Press.

Maudlin, Tim. 2007. *The Metaphysics within Physics*. Oxford: Oxford University Press.

Putnam, Hilary. 1974. "The Meaning of 'Meaning'." In *Language, Mind and Knowledge*, edited by Keith Gunderson, 131–93. Minneapolis: University of Minnesota Press.

Quine, W. V. O. 1969. "Ontological Relativity." In *Ontological Relativity and Other Essays*, 26–68. New York: Columbia University Press.

Rosenhouse, Jason. 2009. *The Monty Hall Problem*. New York: Oxford University Press.

Stewart, Dugald, James Mackintosh, John Playfair, and John Leslie. 1835. *Dissertations on the History of Metaphysical and Ethical, and of Mathematical and Physical Science*. Edinburgh: A & C Black.

Symons, John. 2008. "Intuition and Philosophical Methodology." *Axiomathes* 18: 67–89.

van Fraassen, Bas. 2004. *The Empirical Stance*. New Haven: Yale University Press.

van Heijernoort, Jean, ed. 1967. *From Frege to Godel*. Cambridge, MA: Harvard University Press.

Varzi, Achille. 2011. "Mereology." In *The Stanford Encyclopedia of Philosophy*, edited by Edward N. Zalta. Spring 2011 ed. http://plato.stanford.edu/archives/spr2011/entries/mereology/.

Weatherson, Brian. 2010. "David Lewis." In *The Stanford Encyclopedia of Philosophy*, edited by Edward N. Zalta. Summer 2010 ed. http://plato.stanford.edu/archives/sum2010/entries/david-lewis/.

Williamson, Timothy. 2005. "Armchair Philosophy, Metaphysical Modality, and Counterfactual Thinking." *Proceedings of the Aristotelian Society* 105: 1–23.

{ 15 }

Endogenous Uncertainty and the Dynamics of Constraints

15.1. Introduction

Constraints, once regarded as essential safeguards of a stable social and economic order, have come to be viewed as objectionable on political, economic, and philosophical grounds. They are seen as hindering the pursuit of truth, as violating individual autonomy, as restricting the free and peaceful unfolding of political forces, as leading to inefficiencies in trade and investment, and, last but not least, as obstructing the spread of knowledge. These are powerful reasons against imposing unnecessary constraints on individual and social behavior. Against this backdrop, the persistence of constraints in even the most advanced and enlightened societies must appear as one of the most striking features of the contemporary world. As objectionable as constraints may seem in general, their persistence raises the question whether some of them under certain conditions might benefit society and by what mechanisms such benefits may accrue. These are the questions which I seek to address in this paper.

In premodern times, constraints—unless introduced or abolished in response to a dire emergency—were thought of as ubiquitous and, if not timeless, varying only on timescales exceeding the human lifespan. Modernity, in contrast, is routinely presented as the gradual dismantlement of constraints. In my view, however, modernity, rather than having fewer constraints, is better characterized as being more dynamic, indeed far more dynamic, in the creation, modification, and elimination of constraints. Even as constraints are eliminated in one area of life, new constraints are erected in another and most undergo constant tinkering. We have witnessed the steady erosion of gendered and racial constraints in employment but also the imposition of more severe restrictions on the use of child labor. Due to changes in technology, economic prosperity, and diminished political tensions, constraints on freedom of movement seemed not long ago to be a thing of the past. Recent concerns about terrorism, the spread of disease, and ecological concerns may have reversed that trend. It is natural to associate social dynamics with change and economic freedom with progress. Guided by the same intuition, one is tempted to link constraints with stasis. By highlighting the persistence of constraints, we may even unwittingly reinforce the image of constraints as static features of society. My intention, however, is quite the opposite. I wish to present constraints themselves as constantly evolving, not in isolation but in conjunction with overall social change.

The basic motivation for the research reported here is the desire to identify the conditions under which constraints are either beneficial or detrimental to society. If societal constraints do persist and vary, as I believe, in response to dynamic social and economic conditions rather than being vestigial features of the past, could or should they be viewed as potentially adaptive in nature? And if constraints are among society's adaptive capabilities, to what conditions do they promise a suitable response?

15.2. Endogenous Uncertainty

All economic interaction and coordination requires that agents possess some degree of knowledge about other agents' potential behavior. Agents seeking to improve their lot in an environment with few incentives for cooperation and plenty of room for maneuver without mutual interference will do best when left to exercise total freedom. This self-evident observation underlies the rationality of unfettered individualism at the "frontier." But most frontiers have vanished or shrunk to mere mirages of their former selves. Acknowledging this fact, such champions of freedom as John Stuart Mill attempted to construct a universal and rationally justified notion of minimally restricted freedom by demanding that freedom should extend up to the point where it conflicts with the freedom sought by others. This "golden rule" is prima facie both noble and eminently reasonable except for the fact (recognized by Mill) that most of life, and certainly much of economic life, would grind to a halt if we were to obey it. This was Mandeville's point—driven to pamphleteering extremes—in *The Fable of the Bees* ([1714] 1997) and subsequently conceptualized by others as the basis of laissez-faire theories.

If anything, the overlaps in the spheres of free action have increased substantially in the contemporary "global" world, and the resulting mutual interaction has been thematized by many as interdependency, be it among individuals, firms and institutions, or nation-states. Recognizing this state of affairs is, in my view, necessary lest we lapse into utopian fantasies about either an idealized frontier life or optimized sphere packing. More intensive mutual interaction, in turn, gives rise to what I shall refer to as "endogenous

uncertainty." It is my contention that whereas in the past most instability was
due to exogenous variations—or, at least, variations which were perceived as
having external sources—most of the instability in the modern world is, and
is correctly felt to be, endogenously induced by the behavior of other social
agents (be they individuals or institutions). There is no reason to assume that
the frequency of exogenous shocks has diminished, but the balance between
external and internal sources of uncertainty has shifted in favor of the latter in
a historically short period of time.

In the following, I shall first decompose the notion of "endogenous uncertainty" into key constituent mechanisms within the modeling framework
I have employed. Next, I shall discuss the representation of various types of
constraints. Finally, I shall present and discuss some current results of the investigation on the central question posed: whether, under what circumstances,
and in what manner could constraints be an adaptive mechanism for coping
with the presence and ever changing nature of this type of uncertainty?

15.3. A Landscape of Endogenous Uncertainty

In order to speak meaningfully of endogenous uncertainty in the sense
presented above, there must exist an environment in which individuals
have the same interest, where the pursuit of this interest by one may cause
perturbations capable of hindering others acting on the same interest, and
where such perturbations can propagate through the entire system to the potential detriment of all. Sketching with the broadest brush, the current model
meets these requirements by placing individual agents on a deformable landscape of varying heights representing different levels of individual welfare.
I endow agents with the desire to improve their lot by seeking higher ground
and I introduce mechanisms capable of transforming individual moves into
more or less intense cascades of change affecting the shape of the landscape
and hence the welfare of other agents. At the risk of stretching the metaphor
a little, the four mechanisms of endogenous uncertainty discussed below can
be thought of as four critical dimensions of social interference which can
rarely be avoided in any sufficiently complex society, all attempts to follow the
"golden rule" notwithstanding.

15.3.1. ELASTICITY

The more agents a particular site on the landscape supports, the deeper it will
subside. The response function of each site to varying rates of occupation is
a power curve with exponents sampled from a distribution over a domain of
negative real numbers. As such, both the overall landscape and individual sites
can degrade more or less steeply under increasing occupation. I shall refer to

a site having elasticity of, say, 30% if the welfare value declines by 30% due to a doubling of agents occupying that site. Setting the exponent to the value of 0 for all sites returns the familiar rigid fitness landscape used in many contemporary models.

Deformable landscapes have been investigated previously. Such studies involved landscapes with a single peak shifting randomly to a neighboring position (Nilsson and Snoad 2000), landscapes emulating very slow transformations representing genetic drift (Wilke and Martinetz 1999), and linked coevolutionary landscapes (Kauffman 1993; Ebner, Watson, and Alexander 2000) modeling the interaction of two species. In these cases the cause of deformations is either entirely exogenous (stochastic) or the result of coupling two (or more) subsystems each lacking in endogenous causation. The landscape I employ is different inasmuch as its deformation is truly endogenous. It is capable of producing as well as propagating uncertainty as a result of the agents' own behavior.

The elastic landscape is open to various interpretations. In economic terms it invites comparisons to the supply of and demand for employment of varying degrees of attractiveness. Transferred to the domain of ecology, it may be thought of as a web of habitats which degrade more or less rapidly under intensifying exploitation. The much-discussed tragedy of the commons (Hardin 1968) is a rather contrived case of unsustainable exploitation where a single peak retains its attractiveness over all other sites in spite of its gradual degradation until the final cataclysmic collapse. I can recreate this and other idealized models within my apparatus, but I believe that the presence of multiple welfare peaks and the ability of agents to move from one to another is an essential element of societal dynamics that a sound model must encapsulate. If, as is the case on rigid landscapes, this "moving about" has no consequences on other agents' conditions or prospects, then it is questionable whether one should speak of a society at all and the very notion of endogenous uncertainty becomes vacuous: hence my emphasis on landscape elasticity.

15.3.2. DENSITY

The second parameter is simply the number of agents occupying a landscape with a given number of sites. Increased population density in the modern world requires no further commentary, and general public awareness of its consequences has heightened in recent years. Intensifying migrations resulting in widely varying local changes in density and more pronounced spatial unevenness are not new developments either. In real life, unlike a rigid landscape where density is of no consequence, it is self-evident that varying levels of population pressure—on local and global scales—strongly affect the potential for mutual interference between agents. In combination with elasticity, overall changes in density and local fluctuations become a powerful source of endogenous uncertainty.

15.3.3. CONNECTIVITY

The discrete space underlying our landscape is a regular undirected graph with a toroidal structure. At any given step, agents can execute a single move leading from one vertex to a connected vertex of the graph. The degree of a vertex (i.e., the number of edges originating and ending at that vertex) is a uniform, global parameter k in our model. We shall refer to connectivity as the ratio k / (N – 1), where N is the total number of vertices. Depending on the context chosen, different levels of connectivity can be interpreted as offering a more ample or a narrower opportunity space and greater or lesser potential for social, economic, or geographic mobility. In combination with elasticity, however, connectivity can be viewed more formally as a proxy of spatial interdependence. The commonly heard thesis that as a result of globalization the world has "shrunk" (or even "flattened") is a reflection of the increased likelihood that events at other sites may affect one's welfare. Increased connectivity on an elastic landscape has precisely that effect, providing a third major source of endogenous uncertainty.

15.3.4. INFORMATION DELAY

Uncertainty of any kind has a strong temporal element: it can vanish quickly or endure for longer depending on how quickly information becomes available. Acting on fresh information is more likely to be in one's interest than acting on dated information. The choice between asynchronous or synchronous updating in formulating a multiagent model is a choice between the extremes of agents either remaining fully ignorant about moves made by other agents until the next iteration or receiving instant notification about each and every move. Each extreme is a serious mischaracterization of social life, but this can be safely ignored on a rigid landscape where moves made by others are of no consequence to one's own prospects. My earliest explorations of the properties of the elastic landscape sufficed to convince me of the unsuitability of either extreme and of the need to develop a parameter governing the rate at which information becomes available within successive iterations. To this end, each iteration is subdivided into b batches (or subiterations). Within each batch a random subset of size P/b of the population is selected and allowed to move in a fully synchronous manner. At the completion of the batch, there is a full update and hence the successive batches are asynchronously related. Clearly, b = 1 is the fully synchronous case, whereas b = P represents the other extreme. For all intermediate values, there are, within any iteration, some events that can interfere with each other to produce greater uncertainty and some that are mutually independent. This parameter is the temporal equivalent of connectivity inasmuch as it determines how much temporal separation there must be between two events before one can assume that they are likely not to interfere with each other.

These four components of endogenous uncertainty are at the core of the model. Elasticity is primary inasmuch as complete rigidity renders the remaining three mechanisms toothless and uncertainty vanishes altogether. With any degree of elasticity, uncertainty becomes a nonadditive function of all four parameters. Their respective settings form the parameter space in which we can meaningfully investigate whether constraints of one type or another are likely to foster or hinder the progress of a society of agents. Real life is, naturally, much richer and variegated, and the nature of the functional relation will be system-dependent. Real societies at any time and place also have to cope with more internal mechanisms of consequence, and the precise nature of such mechanisms varies across time and place. I do think, however, that no society can reach any degree of internal organization without acquiring these four components of endogenous uncertainty and that these four in conjoint interaction represent a sufficiently complex environment in which an adaptive process is likely to emerge. I turn now to the social constraints which, in my view, are a key aspect of this adaptive process.

15.4. Mobility, Constraints, and Collective Welfare

I believe that attempts to define the notion of constraints in some absolute sense are highly unproductive. Whether in real life or in a model, constraints must be seen relative to some reference state or condition. In real life, the reference condition may be freedoms granted or withheld in the past, or available to some but not to others in the present, or freedoms which could be realized in the future. In the current model, I introduce constraints as governing the manner in which agents can move with the unconstrained reference case being the prototypical Homo economicus seeking maximal gain.

Maximum Gain agents, the reference agents in our model, move from their current position on the landscape to a connected node with the highest current fitness value (or retain their position if they occupy a local peak). I have so far tested three additional modes of locomotion on the landscape representing increasingly severe constraints relative to Maximum Gain.

Any Gain agents scan neighboring nodes in random sequence but are constrained to terminate the search at the first site encountered which offers an improvement on their present state. In evolutionary models on genetic landscapes, Any Gain is the standard representation of mutation. In the context of the present work, an Any Gain agent can be viewed as acting in a social setting where holding out for maximal benefit is impossible.

Density-Dependent Gain agents are subject to constraints yet more complex and severe. They remain in their current position if it is less crowded than directly accessible sites are on average. If that is not the case, they will act as Any Gain agents on the set (if nonempty) of neighboring nodes which are both higher in fitness and less crowded than their own.

Finally, *No Move* represents the maximally constrained locomotion in which agents are barred from moving at all. I consider it—its triviality notwithstanding—as a base case for calibrating other types of behavior.

The question of how to measure collective welfare—at any given time or dynamically—is, of course, far from settled. It may well remain unsettled, for the choice of a totalizing function is as much a social act as are the events whose effects it seeks to measure. I use four measures. Each relies on taking the mean welfare of the population at every iteration and accumulating these values over the entire experiment.[1] In each case, I normalize the values to eliminate the inherent effects of different parameter choices (greater density and/or elasticity result in a landscape that is more depressed from the start). The four measures result from two binary choices: whether or not to associate a cost with each move on the landscape and whether or not to take into account the time value of benefits. With regards to the former, I use either a zero cost associated with moves or a value derived from the initial state of the landscape in order to guarantee comparability across various parametric settings. The time value of benefits involves discounting future benefits at some "rate of interest." Although the arithmetics of discounting are the same as in a strictly economic setting, my interpretation of it is very different. I take the discounting rate to be a proxy of intergenerational solidarity. The choice of 0% represents idealized societies which value future benefits as much as current ones thereby exhibiting the greatest possible intergenerational solidarity. The choice of 5% represents societies with more pronounced preference for benefits in the near term and lesser concern (rightly or wrongly) for the welfare of future generations.

Having presented the principal parameters of endogenous uncertainty, a brief description of our implementation of constraints, and a discussion of the choice of measures for collective welfare, I can now turn to the obvious question: in what, if any, regions of parameter space do constraints offer benefits at the collective level?

---

[1] It is generally agreed that inequality among individuals and over time must be part of a welfare function. The current model tracks inequality (Gini and Theil indices), but their development is largely invariant to the parameter settings explored in this paper.

15.5. Results

I first compare the performance of two populations, one consisting entirely of Maximum Gain agents and the other entirely of Any Gain agents. An individual run contains 100b updating steps where 1/b, the information delay, is the fraction of the population searching between steps. Each of the cells in the figures represents the mean present value of the welfare outcomes over 100 runs made at the given parameter settings. The collective welfare measure includes a cost for agents to move and a 5% discounting rate for future value. (For details, see the supplementary information section in the appendix). Plate 15.1a then illustrates the results for various population and elasticity levels when the information delay = ½ and the connectivity = 0.1. A red cell indicates that the Maximum Gain population has the higher cumulative population welfare, whereas an orange square indicates superior performance by the Any Gain population. It demonstrates that in regions of high elasticity, the Any Gain population outperforms the Maximum Gain population.

In plate 15.1b, each of the nine blocks is a replica of plate 15.1a, but at different settings for connectivity and information delay. It shows that with values of connectivity of 0.1, 0.2, and 0.4, and values of information delay of ½, ⅓, and ⅙ the dominance of Any Gain in regions of high elasticity is invariant across different values of connectivity and information delay.

In order to appreciate the relative performance of Any Gain, it may help to realize the extent of welfare forgone by adopting this type of search. On a rigid landscape with uniform vertex degree k and neighboring sites ranked in ascending order of welfare value, the Maximum Gain agent will move to the site ranked k, whereas an Any Gain agent will reach only the site ranked 0.75 k on average. Given the order statistics of the Pareto distribution, the "modesty" of Any Gain at k = 30 translates into a welfare value 60% below that of Maximum Gain. (For k = 10, the loss is about 40% and for k = 60 it rises to 70%.) This massive handicap—debilitating on a rigid landscape—does not simply vanish on an elastic landscape (although it may lessen as the order statistics change) but is compensated by a virtue unrequited on the rigid landscape: Any Gain agents spread out and are less likely to bring about a collapse of welfare on an attractive site.

Next, I considered a three-way comparison between the performances of pure populations of Maximum Gain agents, Any Gain agents, and Density-Dependent Gain agents (plate 15.2). As before, a red square indicates that the Maximum Gain population has the greatest cumulative population welfare at that parameter setting; an orange square that the Any Gain population is the winner, and a blue square that the Density-Dependent population does best. A checkered square indicates that the performances of the two populations corresponding to those colors are statistically indistinguishable, and a gray square denotes that none of the three types dominates. Here the results are even more pronounced, in that Density-Dependent agents outperform their less constrained rivals over most of the region of high uncertainty. Maximum Gain naturally retains its advantage in regions of low uncertainty, but Any Gain dominates nowhere.

Because the density-dependent constraint is intuitively counterproductive, the superior performance of that population needs some comment. Although, at root, the density-dependent population benefits from spreading itself over the landscape in a more efficient way, I emphasize that it does this spontaneously. No predetermined "rational" allocation of agent density to sites is involved. In parameter regions corresponding to low uncertainty (high rigidity, low density, etc.), societies can behave as if the landscape were completely rigid, but they will do better to adapt into density-dependent societies as the uncertainty increases. (I remind the reader that the models involve comparisons between pure populations and not competition between subpopulations, so a minority of maximizers cannot exploit a majority density-dependent population.)

Finally, I considered how an extreme case, a population consisting of No Move agents, those who forever remain on their initial nodes, would fare against other populations. I also examined how the choice of measure affects the outcomes. I found that Maximum Gain retains its dominance in areas of high rigidity, whatever outcome measure is used, but that the No Move population becomes dominant as first, a cost for moving is introduced, and then both a moving cost and discounting for future gains. These results are displayed in plate 15.3a.

A full comparison between search methods is made in plate 15.3b, where each of the nine main blocks contains a replica of plate 15.3a for different values of connectivity and information delay. Here we see that the results from the four-way comparison of plate 15.3a remain unaffected by changes in the values of connectivity and information delay. In regions of extreme uncertainty, the population of No Move agents, represented by light blue, outperforms each of the other three types of search procedures. This is because in such regions of extreme volatility, populations of Maximum Gain, Any Gain, and Density-Dependent agents suffer a decline in aggregate population welfare from their initial levels, whereas populations of frozen agents obviously remain at their initial levels. In such periods of extreme and uncertain change, the policy of maintaining the status quo until a more predictable era begins is no doubt a sensible one. Such regions of extreme uncertainty are not descriptive of most areas of contemporary society, and the challenge for societies is to find policies that will push back the regions in which the No Move policy dominates. Plate 15.3b also clearly shows that there are broad regions of parameter values in which the Density-Dependent search procedure does best in the four-way comparison. Thus, in societies with modest amounts of uncertainty and in which agent mobility is highly valued (the majority of nontraditional societies), given the results described in plate 15.3b, Maximum Gain search is to be avoided and a society that is capable of persuading its members to change from Maximum Gain search to Density-Dependent Gain search will be better off. Finally, I note that there are many search types other than the four we have examined which may outperform the No Move policy when great uncertainty is present, while still allowing for some agent mobility. One such possibility is to let a single agent with the lowest fitness adopt the Maximum Gain search; another would be to let several agents sequentially search using the Maximum Gain approach.

15.6. Conclusions

Constraints on individual liberty have a deservedly poor reputation. They are often anachronistic or poorly conceived. Their unintended consequences can become onerous over time, and they are easily abused to serve narrow interests. Yet I can think of no society nor any instance of social progress that lacks a matrix of dynamic constraints. Constraints are familiar, functional, and culturally characteristic features of societies for a reason. When automobiles first appeared, few traffic rules were necessary, but as their density and speeds increased, tighter regulations became appropriate for the enhanced safety of all. Antitrust legislation severely limits the quest for victory in business, but it is generally accepted as essential in ensuring sustained competition and innovation. In these and many other cases the penalties of constraints in terms of costs and loss of freedom are regarded as acceptable or negligible compared to the gain. Beneficial or not, constraints are sociocultural adaptations and as such are salient features of social dynamics.

Our results show that constraints are effective in enhancing the welfare of all under certain conditions of endogenous uncertainty. But what justification is there to view constraints as sociocultural adaptations, in particular, since our model is not adaptive in any explicit sense? After all, we have homogenous populations of immortal agents who cannot reproduce differentially (as in a biological model) to show the superiority of this or that constraint. Nor can they "put the money where their mouth is" (as in game-theoretical models), by punishing those who violate constraints. I have followed neither of these modeling paradigms because I do not believe, nor do I see any evidence, that the dynamics of constraints are biological or game-theoretic in their nature. Their origins are in sociocultural, economic, and political processes which are far too complex for modeling but must remain an essential part of the model's interpretation. What the results do show is that a healthy society equipped with the mechanisms of self-governance will benefit from seeking to tune its own constraints to various degrees of endogenous uncertainty. Finally, I suggest that this tuning is more akin to evolutionary development than to rational policymaking because the effects of introducing, eliminating, or modifying constraints are sufficiently intractable for the decision-making to be political rather than scientific.

At various points I argued that endogenous uncertainty is on the rise. It has been estimated by financial research analysts that the share of stock market volatility due to internally induced fluctuations which self-amplify in spite of the smoothing effects of arbitrage may in recent years have risen as high as 80%. The appearance of greater endogenous uncertainty—whether on Wall Street or on the grander scale of human affairs in general—should come as no surprise. At the core of the program of globalization is the belief that a finer division of labor across greater scales of space and time, and hence more dependencies, will yield greater wealth. What is missing perhaps is the concomitant realization that the resulting increase in endogenous uncertainty may overwhelm the benefits unless we also consider novel forms of constraints.

I hope to have made it clear that my views on constraints are far from Panglossian. As adaptively beneficial as they may be in certain circumstances, most constraints may well be bad and certainly all require constant scrutiny. The reasons are manifold. Adam Smith himself—not known for his advocacy of intervention—pointed out repeatedly that while it is a bad idea to hinder the individual baker in pursuing his quest for greater profit, merchants, as a group, will tend to erect constraints to serve their own narrow interests rather than the overall welfare unless constrained by society from doing so. Thus constraints can be self-serving, while others outlive their overall usefulness but are defended by those who have come to rely on them and who can still derive benefit from them. Others suffer from the law of unintended consequences or its corollary of well-conceived intentions that fail in practice.

The results also indicate that the choice of the outcome measure (the welfare function in economic parlance) is of crucial relevance. Whether or not one applies discounts to future benefits and at what rate can make a decisive difference, even in this simple model. Unless a reasonable level of consensus prevails on what matters to us collectively, the dynamics of constraining and unconstraining are likely to drift rather than to adapt. The rate of discounting, for instance, is not simply a matter of financial facts but a statement about intergenerational relations, which is at the heart of current concerns about sustainable practices. In conclusion, I would argue that, if anything, critical thinking about constraints is needed more than ever. But the critique must be constructive and well informed about the complex nature of constraints in society. The wholesale rejection of constraints is as profound in misreading human history as it is flawed as a policy framework for the future.

Appendix: Technical Information

15.A.1. The Elastic Fitness Landscape

The landscape H is formally described as a 3-tuple:

$$H = \{G[N,c], W, Z\},$$

where $G[N,c]$ is an undirected regular graph of toroidal topology with N vertices, hereafter called nodes, and of uniform vertex degree c, interpreted in the model as the degree of connectivity. The values in W represent the inherent fitness of the nodes: for this model $W = \{w_1, w_2, ..., w_N\}$ is a set of positive real numbers randomly drawn from a Pareto (1, 2.5) distribution. The long tail of the Pareto was truncated at $w = 16$ to avoid extreme outliers producing peaks that would only minimally deform even under extreme loads. The values in Z determine the nodes' degree of elasticity. The model has the ability to assign $Z = \{\zeta_1, \zeta_2, ..., \zeta_N\}$ as a set of positive real numbers randomly drawn from a symmetrical triangular distribution supported by the domain [0,1], but in the present experiments I restricted myself to uniform values of elasticity (i.e., each node has the same degree of elasticity). The effective fitness of a node i at time t is given by

$$f(i,t) = w_i \cdot Max\left[1, \omega(i,t)\right]^{-\zeta_i},$$

where $\omega(i,t)$ is the number of agents occupying node i at time t.

The setting $\zeta_i = 0$ for all i returns a traditional rigid fitness landscape.

The initial occupation of the landscape is arrived at by randomly distributing a population of P agents across the N nodes. The experiments described in this paper set P at values 10, 20, 40, 80, or 160. N is fixed at 125.

Elasticity. I express elasticity as the constant percentage factor L in fitness after a doubling of the number of agents on a node. Formally:

$$L = 100 \cdot \frac{w_i(2 \cdot \omega(i,t))^{-\zeta_i}}{w_i \omega(i,t)^{-\zeta_i}} = 100 \cdot 2^{-\zeta_i}$$

where ζ_i is the value of the parameter drawn from Z. $L = 100\%$ is a rigid node; at $L = 50\%$ fitness drops to half its original value for every doubling of occupying agents. Elasticity is a measure of the impact of other agents' behavior on one's own well-being.

Connectivity. Parametrized by c, connectivity is the spatial measure of endogenous uncertainty. With higher connectivity, there are more neighboring sites whose current occupants may move to a given node and affect its fitness.

I use the term *spatial* in a broad sense of a set with a distance metric which could represent geographic, social, or cultural distance. Greater connectivity substantially modifies the topography of an uncorrelated landscape since the expected number of fitness peaks is inversely proportional to connectivity. The results indicate that the actual number of fitness peaks on landscapes of a given connectivity fluctuates in a manner well described by a Poisson distribution. Given the elasticity of our landscape, peaks will vary in number and their locations.

Chronicity. In many instances of agent-based modeling one must make a principled choice between synchronous and asynchronous updating. Within any temporal iteration (t, t + 1), agents either act in full ignorance of what other agents do in the same period (the synchronous case); or they are fully aware of the effects of the behavior of agents who have preceded them within that period (asynchronous updating). Recognizing that neither extreme properly captures the temporal dimension of social life and that one component of the greater endogenous uncertainty in the modern world is due to the increased degree of synchronic actions in a society, I have implemented a simple method for continuous variation between the two extremes. Each iteration is subdivided into k batches (or subiterations). Within every batch a random subset of size $k^{-1} P$ of the population is selected and allowed to move in fully synchronous manner. At the completion of the batch, there is a full update and hence the successive batches are asynchronously related. Clearly, $k = 1$ is the fully synchronous case, whereas $k = P$ represents the other extreme of complete asynchronicity. For all intermediate values, there are, within any iteration, some events that can interfere with each other to produce greater uncertainty and some that are mutually independent. Chronicity is of no consequence for Highest Gain or Any Gain agents on a rigid landscape, although Density Dependent Gain agents will be affected even on rigid landscapes by the degree of chronicity. With increased elasticity, low k-values may result in all agents on nodes connected to a higher peak simultaneously moving to that location. The formerly attractive peak may collapse under the onslaught, resulting in yet another mass migration to a newly formed peak on the next round. Such self-induced forms of pseudocoordination and the resulting oscillations in fitness do occur in social life (such as social fads and investment bubbles), but they are rare events whose low frequency we can reproduce at intermediate values of k.

Density. By far the easiest parameter to define, it is nevertheless of great consequence in social (and biological) life. It is simply the mean number of agents per node: $P N^{-1}$. Density is the simplest measure for the social equivalent of interdependence and uncertainty whose spatial and temporal aspects I have described above. It is a proxy for how many other agents may interfere with one's best intentions at any given time.

15.A.2. Outcome Measures

I use three classes of measures, each with ample literature and supported by an intuitive understanding of the key dimensions of what it means to live in a healthy society.

15.A.2.1. MEASURES OF PROGRESS IN POPULATION FITNESS

Recalling that the fitness of individual agents is derived from the site they occupy, the mean fitness of the entire population at time t is given by

$$\bar{f}_t = \frac{1}{P} \sum_{i=1}^{N} \omega(i,t) \cdot f(i,t)$$

I have developed four alternative ways of measuring the overall performance at the population level from the first to the last iteration. Each can be derived from the following formula:

$$\Phi(r,C) = \sum_{b=0}^{100} r^{k \cdot b} \cdot \left[(\bar{f}_{k \cdot b} - \bar{f}_0) - m_{k \cdot b} \cdot C \right]$$

The value $\bar{f}_t$ is the mean fitness of the population at time t; r is a discounting factor for determining the net present value of future improvements; the value m_t is the fraction of agents changing position at iteration t, and C is a constant migration cost factor. I index over b, the consecutive batch number (see under "Chronicity" above), and hence $k \cdot b$ is equal to a full iteration. (In other words, I sample the values at the end of each full iteration.) Each simulation consisted of 100 full iterations with 100 k batches.

The mean fitness is normalized as a difference from the initial fitness in order to eliminate the effects of greatly varying initial conditions; i.e., I measure improvement compared to the initial state. The four measures used and their justification are as follows:

- $\Phi(1,0)$: there is no discounting and there are no costs associated with moving from node to node. The absence of discounting reflects a long-term view and a strong notion of intergenerational solidarity since it equates improvements far out in the future with improvements of the same magnitude but within the near future. Assigning no costs whatsoever to migration is defensible in a setting within which agents see variation as routine.

- $\Phi(1,\bar{f}_0)$: no discounting and the cost of migration is set at the mean initial fitness. The latter choice is debatable but recommends itself as the natural method to measure cost in terms equivalent to purchasing power parity. So, for instance, simulations with greater density will result in depressed fitness values compared to simulations with lower density. There is no common currency between two such worlds, and introducing a constant cost independent of initial conditions will result in severe distortions.

- $\Phi(0.95,0)$: no cost to migration but future improvements are discounted to the present at a rate of 5% per iteration. This is in line with the general notion of the time value of money and reflects a moderate preference for improvements earlier rather than later.

- $\Phi(0.95,\overline{f_0})$: the combination of both discounting and costing. It is the measure that we have used in our work.

Note that all of these measures rely on mean values obtained at the level of the entire population.

Acknowledgments

Much of the research described here was done in cooperation with Tiha von Ghyczy of the Darden Graduate School of Business, University of Virginia, and many of the ideas are due to him, including the central device of a deformable landscape. Thanks are also due to Janis Antonovics, Walter Fontana, William Macready, Martin Reeves, and Douglas Taylor for comments on earlier drafts. Partial support for research was provided by National Science Foundation award SBE0523678 and by the Centre National de la Recherche Scientifique, France.

References

Ebner, Marc, Richard A. Watson, and Jason Alexander. 2000. "Co-evolutionary Dynamics on a Deformable Landscape." In *Proceedings of the 2000 Congress on Evolutionary Computation*, vol. 2, pp. 1284–91. Piscataway, NJ: IEEE Press.

Hardin, Garrett. 1968. "The Tragedy of the Commons." *Science* 162: 1243–48.

Kauffman, Stuart A. 1993. *The Origins of Order*. Oxford: Oxford University Press.

Mandeville, Bernard. (1714) 1997. *The Fable of the Bees: And Other Writings*. Abridged and edited by E. J. Hundert. Indianapolis: Hackett.

Nilsson, Martin, and Nigel Snoad. 2000. "Error Thresholds for Quasi-Species on Dynamic Fitness Landscapes." *Physical Review Letters* 84: 191–94.

Wilke, Claus O., and Thomas Martinetz. 1999. "Adaptive Walks on Time-Dependent Fitness Landscapes." *Physical Review E, Statistical, Nonlinear, and Soft Matter Physics* 60: 2154–59.

Explanation, Understanding, Ontology, and Social Dynamics

Chapter 12 is the oldest paper included in the collection. Having been unable to solve for many months a problem that is peculiar to probabilistic causality, I saw the solution suddenly while sitting at my desk.[1] The problem in question was what to do with properties that lowered the probability of the effect. Should they be considered on a par with factors that increased the probability, did they play a different and distinctive role, or should they just be ignored? Wesley Salmon (1984) had argued persuasively for the first option, but that approach seemed to me not to capture what was distinctive about probabilistic causation. The solution developed in chapter 12 separates contributing and counteracting causes, allows the value of the probability to fall out naturally as a result of including all probabilistically relevant factors in the conditioning events, and in so doing, denies the third option above.[2]

Probabilistic causality can, in my view, be properly understood only within a single-case propensity account of probability, and it applies in a nondegenerate way only within indeterministic systems. That does not mean chance should be reified within such accounts, for chance does not exist, in any familiar sense of existence. The reasons for this are straightforward. Probability theory had dual origins in finite relative frequencies and in an informal type of epistemic uncertainty. Neither of those approaches requires adopting a realist interpretation of chance. For the frequentist Richard von Mises, the reason was his deep commitment to positivism, together with the fact that he was forced to introduce idealizations about relative frequencies into his mature theory

---

[1] I did not, following Russell's example, exclaim, "Great God in boots, the argument is sound!" Russell's account of the moment he realized the ontological argument was sound is given in 1967, 63. Although in his autobiography Russell's imagination occasionally got the better of him, this report has some evidence on its side (see Spadoni 1976).

[2] The basic theme in chapter 12 is further developed in Eells 2008.

(von Mises 1964). Similar reasons held for Hans Reichenbach's frequentist theory. Kolmogorov went much further in the direction of abstraction and explicitly disavowed any empirical interpretation of probabilities within his nonelementary theory (Kolmogorov 1933, 15). Sociologically, the reluctance to accept indeterminism in the early twentieth century coupled with the significant contributions made to modern probability by Soviet mathematicians who were operating within an official Marxist ideology committed to determinism meant that appeals to chance played a minimal role in shaping the modern theory. Even C. S. Peirce's account of tychism did not play a significant role in philosophical discussions of probability until Karl Popper's propensity account was introduced in 1957. Within that propensity tradition, it is possible to maintain a minimalist commitment to natural chances by describing the structure of the system possessing the chances, then identifying the variables and parameters that affect those chances and, having fixed them, acknowledging that there is nothing left to specify about the system except to state the resulting propensity value. There is no need to introduce chances as an extra ontological feature of the world. An account of probabilistic causation based on such an approach to propensities is a pure case of what Hume taught us to see about causation.

An elaboration of one point may be helpful. Chapter 12 was written before I had realized the importance of Mill's invariance condition for causes; this condition requires that a cause act invariantly in all contexts of occurrence. This is a severe condition that makes for complex causal statements, but the reasons given by Mill and adapted for the probabilistic case (Humphreys 1989, sec. 25) make a convincing case for adopting this condition. It might be objected that Mill's condition cannot be satisfied because of the presence of ceteris paribus conditions, the reason being that ceteris paribus conditions are said to be essentially vague or that they cannot be completely specified. What is needed to address the objection, both for single-case probabilistic causal claims and for general causal claims, is to add "and nothing else is relevant" to the statement of the causal claim. That clause applies whether X occurs in a closed or an open system, but there is a logical issue that needs to be addressed. In the case of binary valued properties, "Nothing else is relevant" can be represented as $\forall F(F \notin \{X, Y\} \to Pr(X|YF) = Pr(X|Y{\sim}F))$, where X is the effect variable and Y is a usually complex collection of causally relevant variables.[3] What is the scope of quantification in this claim? It is genuinely universal in the sense that the universal claim is unrestricted over all variables representing natural properties. There is no need for sortal quantification and the claim is falsifiable, modulo the limitations of statistical hypothesis testing.

---

[3] Pr represents a propensity, not a relative frequency, a degree of belief, or an inferential probability, so principle CI of Part III in this collection applies.

Even if there is an infinite number of natural properties, the quantification covers them all. Universal quantification over all properties is certainly problematical, but I do not know of a conclusive reason for not doing so in this case. Of course the claim may be false, and indeed often will be, but this is an inescapable aspect of any inductively established generalization, and it is not peculiar to the ceteris paribus situation. As a result, there will always be epistemological uncertainty about whether the unqualified generalization applies to a situation (see Humphreys 1989, sec. 41B), but this does not lead to a heterogenous, unbounded, disjunction of possible exceptions, any more than multiple realizability entails a heterogeneous disjunction of realizers (see Humphreys 2016, sec. 6.4). The situation is much more complicated with continuously valued properties, and I do not have a definitive solution for that case.

Explanatory reduction and its failure is a theme that runs through Part II of this collection. Unity of science claims tend to be only indirectly about ontological reduction, the subject being addressed through the apparatus of linguistically formulated theories. Theoretical reduction, whether in the Nagelian, Kemeny-Oppenheim, functional, or some other approach, has traditionally been hierarchical: biology to molecular biology, molecular biology to organic chemistry, organic chemistry to physical chemistry, physical chemistry to physics, and so on. Reductive explanations have usually accompanied such reductive projects; here I shall instead address reductive understanding. The connections between reduction and understanding are complex.[4] It is now widely recognized that there are different forms of explanation that can lead to understanding—causal, unificatory, mathematical, functional, and other kinds of explanation, and efforts to fit these into a common framework have not been successful. There is a reason for this. Humans have different modes of understanding that cannot easily be separated from psychological capacities. These capacities can be associated with different kinds of cognitive abilities, and the degree of understanding gained in a given mode of understanding is often dependent upon the type of representation used. Use an algebraic representation for how three variables covary and for those without good mathematical skills it will produce only limited understanding; use a three-dimensional dynamic graphic and much greater understanding will generally result.

In the case of mathematical understanding, proofs have multiple functions. They can justify the theorem or lemma, they can assist understanding of why the result holds, and they can embed the theorem in the wider corpus of mathematical results so that it can be used as the basis of further results and provide a broader understanding than when the proof is taken in isolation. The relevance relation in the pragmatic account of scientific explanation (as in van Fraassen 1980, chap. 5) can be seen as a way of capturing the different modes

---

[4] For one comprehensive treatment of understanding, see de Regt 2017.

of understanding, but its values can reflect objective features of the situation rather than simply pragmatic interests. Unless there is an incontrovertible reason to the contrary, pragmatics should always be a secondary feature of an epistemological enterprise and not a primary feature. Inverting that priority is an invitation to remain at the level of psychological and linguistic phenomena rather than to explore the ontological complexities of a subject.

One aspect of chapter 13 that illustrates the benefits of keeping pragmatics as a secondary feature of explanations is an argument that those seeking an explanation should not be allowed to set the terms in which the explanation is given. More directly, the presupposition underlying a why-question will often be incorrect due to the questioner's lack of knowledge, and that presupposition will have to be corrected by the answerer. Another situation in which the questioner should not be permitted to fix the terms of the answer is in cases of emergence, where the vocabulary in terms of which the answer needs to be given is different from the vocabulary in terms of which the explanandum is framed, and the questioner may not be familiar with the variant modes of description.

In retrospect, the question raised in section 13.7 strikes me as increasingly in need of an answer. One correction needs to be made, though. I claimed that the aspirin example is different from the cholera example. Although that was correct about the nineteenth-century scientists, the situation of the ancient Greeks was the same as that of John Snow. Willow leaves and willow bark are the carriers of salicylic acid just as water was the carrier of the vibrio cholerae bacteria. The more important issue is the ontological situation. I wrote, "Koch discovered the cause of cholera. This is misleading, for current evidence suggests that it is a toxin manufactured by the *Vibrio cholerae* bacterium that is actually the cause of the disease. Such cases raise an obvious question: Are the only causes those factors for which no further decomposition into relevant and irrelevant components is possible? My answer is, tentatively, 'Yes.'" I would now remove the modifier "tentatively." Despite the common and reasonable attribution of causal discovery to carriers plus causal agents, the ability to materially separate carrier from agent shows the causal irrelevancy of the former. This allows for causally relevant catalytic agents and cases in which the causal agent needs a particular type of environment in which to operate effectively. This has the consequence, doubtless unpalatable to many, that very few of our current causal claims are correct, and not because of theory changes but for reasons of incomplete specification of the actual cause.

A comment is also necessary on the objection that if we require every member of an explanans to be true, we run into the pessimistic induction and this will undermine confidence in current explanations. The pessimistic induction is not grounds for skepticism in the way that classic skeptical arguments such as the evil demon are grounds for doubt. The latter provide an alternative explanation for the facts, and that explanation cannot, given any amount of experience, be eliminated. With the pessimistic induction, we can often in ret-rospect have an explanation for why the abandoned theoretical apparatus was successful even though we now know that it was false. Consider the Ptolemaic theory used as an example in chapter 13. We now know that the reason why this theory (which was primarily a method) was predictively successful is that any continuous orbit can be reproduced by Ptolemaic methods.[5] From our retrospective position of having superior mathematics, we can see why the predictive success did not depend upon truth.

Because the pessimistic induction is an argument against scientific realism, eliminating alternative explanations for the empirical facts is an important part of adopting a realist position with respect to a given theory. Identifying ways in which predictive success can be achieved independently of truth also provides ways to refine a selective realist position. This should be part of a general approach to the pessimistic induction because the proper use of induc-tive inferences requires specifying conditions the satisfaction of which makes for the successful use of the theory at hand. Not all cases of successful theories come with such conditions, and one role of construction assumptions and cor-rection sets (see Humphreys 2004, secs 3.6–3.7) is to specify those conditions in advance. Equipped with the construction assumptions, we can know that the model is false even though it is successful at the time of use. When we can know the falsity only retrospectively, this need not involve a fallacious appeal to the future by the realist. What it does is to remove from our current induc-tive base theories that at the time were thought to be true but are now thought to be false. That is, if we can point to some identifiable difference between the successful and true theories and the successful but false theories, we can use those differences to undermine the extrapolation from historical examples to the conclusion that there will be many successful but false theories in the fu-ture to which realists must be committed.

Chapter 14 argues against a certain way of pursuing ontology that I called "speculative ontology." There is no need to elaborate on the arguments against speculative ontology, which speak for themselves, but a few words on the con-structive position I called scientific ontology would be helpful. Scientific on-tology need not be the ontology to which scientists currently subscribe. With some exceptions, scientists do not spend time critically evaluating the default ontology for their field. They are not trained to do so and have more pressing professional concerns. So deferring to the consensus view in a scientific field because it is the consensus view is unreasonable. It may be argued that philo-sophical ontology should be concerned only or primarily with a fundamental

---

[5] Herget 1939, 311 notes that with a little more flexibility the ancient Greeks could have discovered Fourier series. We should not exaggerate the sophistication of the Ptolemaic system, however. Lakatos 1980, 171 notes that Owen Gingrich's calculations showed that the compilers of the Alfonsine Tables used only a single epicycle.

ontology and not with entities that result from that fundamental ontology. That view is derivative from the kinds of reductive or quasi-reductive positions that are shown to be false in Part II, and we should not suppose as a result that all of our ontological commitments should be a close approximation to those of fundamental physics. The ontologies of fundamental physics and astrophysics are frequently conceptually alien to humans and sometimes highly speculative, often radically so, whereas the ontologies of some less fundamental sciences are stable and epistemically more accessible. It is no slight to physicists working on quantum gravity to point out that knowledge in that area is less secure and successful than are the items of knowledge that doxycycline is currently a safe and effective treatment for Rocky Mountain spotted fever and that its mechanism of action is part of the biochemical domain.

Scientific ontology has a number of components that involve the interplay between ontology and epistemology. The first is that the constraints imposed on scientific knowledge by empiricism are far too restrictive for articulating an ontology that captures the powerful methods that distinguish science from everyday modes of investigation. In the case of vision, the fact that the human visual system uses only the 400- to 700-nanometer part of the electromagnetic spectrum, approximately 0.0035% of the total range of electromagnetic waves, is highly contingent. That it does so is a joint consequence of the fact that most of our sun's energy which reaches the earth's surface lies in that range and a set of evolutionary conditions that made any use of adjoining regions largely unnecessary for humans.[6] These conditions were contingent, but that fact does not by itself undermine empiricism. What does undermine it is the claim that, as a source of evidence, direct perceptual experience is uniformly more reliable than, or at least as reliable as, alternative, artificial modalities. Once one recognizes that this claim is false and can be shown to be false using epistemic criteria that are acceptable to empiricists, we should also recognize that those alternative modalities have different degrees of reliability with respect to what they reveal about the world. One attractive position thus combines what was admirable about empiricism—its insistence that our claims to knowledge be both constrained and driven by contact with the natural world—with a form of selective realism.[7] This is one form of scientific ontology that is attractive.

Selective realism recognizes that realist claims must be tied to epistemic accessibility, that a uniform commitment to all terms in a theory or model having a genuine reference is at odds with the use of idealizations and apparatuses of mathematical convenience in models and theories, that different types of property are differentially accessible to different types of epistemic agent, and

---

[6] For one such explanation see Osorio and Vorobyev 1996. There are others.

[7] Selective realism can be developed in different ways, and detailed formulations of different kinds are given in Chakravartty 2017 and Peters 2014.

that this accessibility is only loosely associated with spatial scales. Selective realism also recognizes that a blanket realist claim that the world is independent of our representations, mental or otherwise, is too simple. We can and do affect the world, and what the world contains is, to a certain extent, dependent upon the representations we use to intervene in it. The discovery of penicillin may have been fortuitous, but the subsequent development of antibiotics and their effects on bacteria has been mostly dependent upon increasingly sophisticated scientific representations of the "unobservable" world.[8] Other, macroscopic, parts of the world, such as cars, would not exist if we (or cognitive agents roughly like us) and our representations did not exist.[9]

Chapter 15 was written prior to the economic collapse of 2008 in the period when the suboptimality of minimally constrained markets was less widely recognized. One motivation for developing the model was the recognition that the economies of advanced technological societies are highly dynamic and, perhaps as a consequence, increasingly unpredictable. In such circumstances, a commitment to inflexible economic or political dogmas of any kind is unlikely to serve society well. The core of the project was to explore whether specific types of constraints help or hinder the maximization of group utility. To pursue this, I had to develop a different kind of representational device. Some type of fitness landscape was a natural choice, but standard fitness landscapes are rigid and do not reflect the effects of the agents on their environment. An important innovation, due to my Darden School colleague Tiha von Ghyczy, was the use of deformable fitness landscapes that capture endogenous effects. Although there had previously been a few simple examples of non-rigid landscapes, none allowed the complex behaviors to which simulations are ideally suited. Allowing the effects of individual maximizing to affect, often negatively, the utilities of the agents and their neighbors introduces sometimes radical, but also realistic, consequences to aggregated individual acts. The simulation was programmed in Mathematica and can be scaled up. The numerical outputs were represented in displays that captured as much information as possible in a two dimensional graphic.

As with many models, the artificial society is a greatly simplified version of real societies, and the goal was to exhibit qualitative effects that were robust across omitted details rather than to make quantitative predictions. For philosophers, perhaps the most immediate point of interest is the significant generalization of the tragedy of the commons scenario that the model allows.

---

[8] I have put the term "unobservable" in scare quotes to signal a shift in its use away from the empiricist tradition.

[9] Claims such as this are implicitly prefixed by the probabilistic hedge "almost certainly," which I have omitted on stylistic grounds.

In particular, rather than a single asset, access to which might be controlled by pricing incentives, the model offers insights into how an element of moderation in utility maximization can provide superior results when multiple options are available to agents.

One limitation of this research was that the model societies do not contain mixtures of agent types; all comparisons are between results from pure populations of a single type of agent. Because of this, the simultaneous use of different search strategies by different types of agents, which is a common cause of conflict, is not addressed, although that case can be covered by an extension of the current model. A further restriction is that during any given run, the number of agents remains constant, so there is no reproduction, death, or leaving or entering the society. The model is therefore best suited to modeling societies with stable population levels that are reasonably homogeneous in the sense that an exiting member is, on average, replaced with a behaviorally indistinguishable equivalent individual.

Recognizing that externally imposed constraints on actions often have negative effects, I was interested in self-constraining systems. The underlying idea is that constraints emerge as a result of the agents' activities and affect the way in which they can act. These evolving constraints both result from and result in a reorganization of the society to which agents belong.

Feedback mechanisms are a form of self-generated constraint but often do not result in self-organization. Four examples of self-constraining systems that are not self-organizing are self-avoiding random walks, drawing from an urn without replacement, the self-limiting spread of an invasive language to saturation, and, less obviously, free speech. In the last case, society, through extensive use of free speech, has found it preferable, through further exercise, not to allow completely unrestricted freedom of expression but to impose some constraints on it. This is not to say that curtailing speech cannot be done through other means—that is obvious—but that the process is capable of constraining itself in ways to which authoritarian societies do not have access.

References

Chakravartty, Anjan. 2017. *Scientific Ontology*. New York: Oxford University Press.

De Regt, Henk. 2017. *Understanding Scientific Understanding*. New York: Oxford University Press.

Eells, Ellery. 2008. *Probabilistic Causality*. Cambridge: Cambridge University Press.

Herget, Paul. 1939. "Planetary Motion and Lambert's Theorem." *Popular Astronomy* 47: 310–13.

Humphreys, Paul. 1989. *The Chances of Explanation*. Princeton, NJ: Princeton University Press.

———. 2004. *Extending Ourselves: Computational Science, Empiricism, and Scientific Method.* New York: Oxford University Press.

———. 2016. *Emergence.* New York: Oxford University Press.

Kolmogorov, Andrei. 1933. *Grundbegriffe der Wahrscheinlichkeitsrechnung.* Berlin: J. Springer.

Lakatos, Imre. 1980. *The Methodology of Scientific Research Programmes.* Vol. 1. Cambridge: Cambridge University Press.

Osorio, D., and M. Vorobyev. 1996. "Colour Vision as an Adaptation to Frugivory in Primates." *Proceedings of the Royal Society of London B: Biological Sciences* 263 (1370): 593–99.

Peters, Dean. 2014. "What Elements of Successful Scientific Theories Are the Correct Targets for 'Selective' Scientific Realism?" *Philosophy of Science* 81: 377–97.

Popper, Karl R. 1957. "The Propensity Interpretation of the Calculus of Probability, and the Quantum Theory." In *Observation and Interpretation*, edited by Stephan Körner, 65–70. London: Butterworths.

Russell, Bertrand. 1967. *The Autobiography of Bertrand Russell.* Vol. 1. New York: Little, Brown.

Salmon, Wesley C. 1984. *Scientific Explanation and the Causal Structure of the World.* Princeton, NJ: Princeton University Press.

Spadoni, Carl. 1976. "Great God in Boots—the Ontological Argument is Sound!" *Russell: The Journal of Bertrand Russell Studies* 23–24: 37–41. http://dx.doi.org/10.15173/russell. v0i2.1462.

van Fraassen, Bas. 1980. *The Scientific Image.* Oxford: Clarendon Press.

von Mises, Richard. 1964. *The Mathematical Theory of Probability and Statistics.* Edited by Hilda Geiringer. New York: Academic Press.

{INDEX}